Eaters of the Dry Season

Eaters of the Dry Season

Circular Labor Migration in the West African Sahel

David Rain

Westview Press
A Member of the Perseus Books Group

Copyright © 1999 by Westview Press, A Member of the Perseus Books Group

Published in 1999 in the United States of America by Westview Press, 5500 Central Avenue, Boulder, Colorado 80301-2877, and in the United Kingdom by Westview Press, 12 Hid's Copse Road, Cumnor Hill, Oxford OX2 9JJ

Library of Congress Cataloging-in-Publication Data
Rain, David.
  Eaters of the dry season : circular labor migration in the West
African Sahel / David Rain.
     p. cm.
  Includes bibliographical references and index.
  ISBN 0-8133-3616-3
  1. Migration, Internal—Sahel.   2. Sahel—Economic conditions.
3. Droughts—Sahel.   I. Title.
HB2125.55.R35   1999
331.12'7966—dc21   99-21974
                                                                         CIP

The paper used in this publication meets the requirements of the American National Standard for Permanence of Paper for Printed Library Materials Z39.48-1984.

10   9   8   7   6   5   4   3   2   1

For Anna

Contents

Tables and Illustrations

Figures

Photos

Preface

Eaters of the Dry Season deals with the seasonal mobility of people along the desert margin of West Africa. The subjects are farmers and herders, traders and beggars, men and women—all of whom share the habit of circulating between their villages of origin and a small city on the border between Niger and Nigeria.

To research this subject, the author did something outrageous: He consulted the migrants themselves and sought their views on the process they actively participated in. The migrants' stories and opinions were instructive. Some felt that given the region's history and marginal recent development in a climate that has always been capricious, it makes no sense to stay put and make a living solely off the land. Others bemoaned the erosion of rural self-sufficiency and the challenge that circular mobility presented to traditional rural values. Many praised the lifestyle that mobility makes possible: It allows individuals and families to spread interests and risks around and to capitalize on enduring social ties. Cultures of the region rejoice in the social position of guesthood that welcomes visitors and keeps the psychological costs of movement low. In times of drought and famine, having a network of family and friends is as necessary as having money or a ticket out. Understanding Sahelian practices provides a window into how people deal with risk; it also provides lessons on how to be mobile and yet still maintain ties to what is important.

The study is contextual and panoramic; it is centered on the immediacy of daily and seasonal routines. It delves deeper than the alarmist or excessively journalistic accounts of African drought and famine that are all too common in the literature. A solid treatment of an overlooked subject, this book presents a broad and evenhanded analysis of the changing relations between urban and rural realms, an analysis that readers will find useful in a variety of settings. Readers will see that mobility is a rational response to uncertainty and that it is far older and more deeply ingrained in the human spirit than current technology leads us to believe. This work will challenge laypeople as well as scholars and policymakers to consider how actual people respond to global changes in the next century.

When outsiders think about the Sahel at all, they often see it as a lost cause, a region where inevitable collisions between population growth and environmental decline have run their course. Thus, the Sahel has become one of the most deterministically rendered regions in the world, with little chance to counter the images of the drought and starvation that are attributed to the region. How much those living in the Sahel can improve their lives depends on the degree to which a massive population realignment toward urbanization takes place. This realignment will accelerate in the next century and create human landscapes that hardly resemble the rural images most people have of Africa. Some assume that the changes will be swift and massive. In fact, change has been happening as long as there have been people there. The surprise is how invisible and incremental this process has been thus far; outside observers are blind to its subtleties.

Changes in population mobility form a component of the urbanization process. Circular mobility—going away in the off-season to seek food or cash-making opportunities—is a very old practice that predates colonialism in the interior of West Africa. Circular mobility is usually neglected because movements that demographers classify under the rubric of "internal migration" in standard typologies are considered less consequential than more permanent movements across international boundaries. When trying to understand why people move, it is necessary to transcend cartoonish depictions of "pushes and pulls" and to consider the actual circumstances of the movers. The reasons for movement are as varied as the people, and we need to leave behind statistics and strive to understand what motivates the movers themselves.

Adapting to the environment has been a hallmark of life in the Sahel since humans first walked the area millennia ago. Surviving in the Sahel requires strategies of low place-based investments, portable livelihoods, economic interests that are spatially dispersed, and, most broadly, the use of mobility to exploit spatial variation in precipitation and natural resources. Yet despite the rationality of this culture of movement, development strategies have been informed by the idea that persuading peasants to stay put and to invest in their villages better prepares them against drought and famine.

The study on which *Eaters of the Dry Season* is based uses a combination of survey-interviews, participant-observation, and geographical analysis to set a variety of circular movements into context. My goal is to illuminate real situations faced by individuals as they move through their everyday lives. I examine the settlement history of the region, population dynamics, changes in farming and land-use practices, and the spread of market relations. These are the realities that have contributed to the popularity of circular mobility in the region.

1

The Changing World of *Cin Rani*

Every year after the rainy season ends and crops are harvested, hundreds of thousands of men and women leave their villages to work in West African cities. In Maradi, a provincial capital near the border of Niger and Nigeria, the seasonal flux of peasants swells urban markets and neighborhoods. After their arrival in the city, peasant workers find employment in day labor, apprenticeships, marketing, or crafts, earning the cash that is funneled back to their home villages to buy food and other necessities. These "eaters of the dry season"[1] are practitioners of traditional dry-season circular mobility.

Circular labor mobility is an economic activity with a long history in West Africa. In the drylands south of the Sahara Desert, known as the Sahel (meaning "shore" in Arabic), seasonal or subseasonal movements across rainfall gradients are a livelihood strategy to maximize investments of time and other resources. Population movements are one response to the variations in the environment and political economy that date from precolonial to postcolonial times. The seasonal circulation of people has provided a reliable hedge against the frequent droughts that plague the region.

Known as the commercial capital of Niger, the city of Maradi is located near the southern border of Niger and is the gateway to Nigeria, fifty kilometers to the south (see Figure 1.1). Maradi has an active twice-weekly market selling goods smuggled over the border from Nigeria, and its intricate Hausa social structure is based on the values of Islam and merchant capitalism. The city absorbs large numbers of peasants, mainly Hausa agriculturalists but also Tuareg, Bougajé, and Fulani herders or pastoralists, into its expanding economy. In 1995, the Mayor's office estimated that Maradi receives 20,000–25,000 temporary circular migrants in a typical year and that it briefly hosts many more who pass

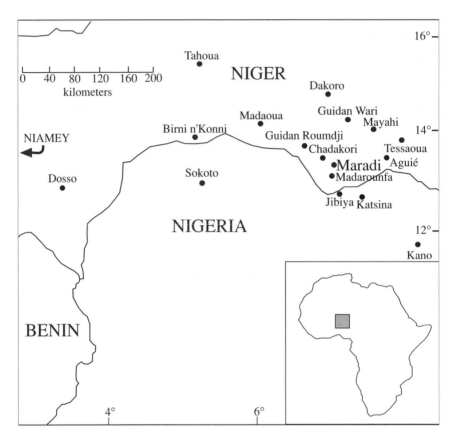

FIGURE 1.1   *Niger-Nigeria Regional Map*

through Maradi en route to such northern Nigerian cities as Jibiya, Kano, and Katsina.

   This book explores the changing geography of seasonal circular mobility in Maradi through a contextual investigation of its practitioners. In doing so, it aims to answer large questions about the nature of change in Sahelian West Africa. How and why did people come to be mobile in this part of Africa? How is population mobility affected by changes in population, economics, and environment in the region? What is the role of mobility in fueling urban growth and rural land-use change? How do individual, household, and community mobility decisions relate to changes in the region, and how do they relate to cultural norms and social networks? How does mobility link the city with rural villages in West Africa? How is it creating divisions within those villages?

Defined simply as people changing location, population mobility is a basic human activity, undertaken by individuals for their own objectives. In most mobility typologies, the term embraces *migration,* which implies the severing of roots in one place and their establishment in another, a practice normally associated with the Western world. It also includes *circulation,* which involves periodic changes in residence following some kind of cycle. The influence of mobility upon people and places is substantial, and yet the concept of mobility remains elusive. It challenges notions dear to demographers about residential permanence and the separation of urban and rural, and it demonstrates the real difficulty of capturing a moving population. Embracing both ecology and economics, it cannot fit into a convenient agenda for development work. Yet it structures how people work, how they see their land and country, and also how they deal with risk.

Demographers normally assume that people base location decisions on an economic calculus and that people keep a fixed residence until they make a permanent move. Since E. G. Ravenstein's "laws of migration" were defined over a century ago, migration has been conceptualized as determined by "pushes" and "pulls" in the origins and destinations. Yet the Western emphasis on permanent moves for job relocations is not relevant everywhere. The laws of migration explain behavior in Western Europe and North America better than in sub-Saharan Africa. Population circulation—though a well-known feature of mobility typologies and of mobility systems in developing countries—is grossly neglected in the theoretical literature. Circulatory movements are characterized by people moving from their places of residence for varying periods of time but ultimately returning to them. The reality of multiple, temporary, circular moves vectoring out from a permanent base has implications that contrast vividly with the decisive "move" in standard migration conceptions. Whether by accident or design, circular mobility is hidden in the shadow cast by assumptions of Western-style residential permanence. This is a predicament only partially explained by the difficulty of recording movements.

There is a conspicuous need for new geographical methods to analyze circular forms of mobility. To join mobility and production geographically at the regional scale, several different analytical yardsticks can be employed. One yardstick can measure how individuals move to satisfy household or community needs, another can gauge how settlements respond to the prevalent mobilities of their inhabitants, and still another can assess how larger cultural systems create and transmit values related to the different mobility types. Although explanations such as ethnicity, drought or environmental change, mode of production, and the needs of

such larger macrolevel forces as the world economic system are all un-
dercurrents to discussions here, each alone is an inadequate explanation
for change. All must be considered at once as they interact with other explanations on the level of lived experience.

Mobility provides a window into how moving people perceive the
places where they live, and it connects economic activities with the behavioral, architectural, and geographical settings where they occur. Understanding mobility requires an inquiry guided by a philosophy of
shared understanding. Such an inquiry can be called a "grounded" geography of human population. It focuses on the particular circumstances of
the movers and the communities affected by mobility, and it considers
the critical roles of place, culture, and the environment in the dynamics of
human livelihoods and reproduction. The chronology and conditions of
how regions are settled, the ways that people perceive and interpret their
surroundings and the world, and how production practices, the satisfaction of food needs, and the mitigation of hazards stem from attributes of
the local ecosystem—or from the relations with outside entities—are all
foci of this subdiscipline.

In West Africa seasonal labor mobility is a very old activity that has en-
dured the changes of the colonial and postcolonial periods and continues
to be a rational response to uncertainty for many people. The practice of
mobility is deeply rooted in Hausa culture, which is based historically on
long-distance caravan trade, hunting, and herding activities. Throughout
Africa, it is natural to move. *Cin rani* and other movements that are partially determined by the climate are common here, for the environment in
many places does not permit sedentary living. Eating the dry season requires a network of connections that spreads across space; it also requires
being quick and resourceful; it requires being a good navigator. The mas-
tery exhibited by the migrants, of eating places, eating seasons, even eating the whole world—in other words, eating things that are thought to be
intangible (not to mention inedible)—is a motif that will accompany us
throughout this book. To the migrants I met and talked with, moving is
as commonplace as eating; those who move are no more remarkable than
those who eat.

And yet as we approach the new millennium, population mobility is
being painted as something wholly new, a response to modernity, capitalism, environmental ruin, or overpopulation. The millennial visions of
the future presented by authors such as Thomas Homer-Dixon (1991),
Paul Kennedy (1993), Robert Kaplan (1994, 1996), and Samuel Huntington (1996) reveal in particular an image of the poor as desperate. The
poor *move*, fleeing areas of conflict or environmental stress and taking up
residence in temporary accommodations in new lands. In their movements, they function as emblems of future disorder. The shantytowns on

the edges of newly industrialized cities where recent urban migrants gather in search of temporary work are portrayed as symbols of failed policies or as breeding grounds for sexually transmitted diseases. The mass movement of refugees personifies the human costs of ethnic conflict. To the United Nations High Commission on Refugees (UNHCR 1995), poor people on the move are the face of want in the latter half of the twentieth century.

Commentators warn that we are entering a world where the plight of millions[2] of refugees and internally displaced persons around the world appears to be continually worsening. William Wood at the U.S. State Department predicts that growing economic and demographic disparities between rich and poor countries are "global trends that will drive most international migration flows to unprecedented levels" (Wood 1994, p. 203). Of particular interest to the international development community are "environmental refugees," that is, displaced people driven out of their customary homes by a smorgasbord of ills including environmental degradation, deforestation, desertification, extreme weather events resulting from weather patterns like El Niño, saltwater intrusion, or prolonged droughts (El-Hinnawi 1985, Jacobson 1988, Suhrke 1991, Myers 1993). Norman Myers envisions 150 million to 300 million environmental refugees and internally displaced persons searching for a haven by the middle of the next century. In a Climate Institute study, Director Mustapha Tolba of the United Nations Environment Program (UNEP) states that "the options for the poor are stark: either to flee, or to stay put and starve" (UNEP 1993, p. 4).[3]

Commentators argue that in sub-Saharan Africa population movements that stem directly from worsening economic and environmental conditions are likely to increase in coming years. Some have argued that over the long term, Africa's capacity to find a path to economic development depends on the degree to which a *necessary* population redistribution takes place (Cour 1994). This redistribution is occurring from rural to urban sectors within countries and from sparsely settled but fast-growing landlocked countries to urbanizing coastal economies. Extrapolating the growth rates of individual African countries provides sobering evidence for the assertion that African mobility may be beginning rather than ending. According to Club du Sahel, the entire period of rapid demographic growth in West Africa will be marked by a large-scale redistribution of the regional population (Club du Sahel 1995).[4]

The personal circumstances of migrants in West Africa epitomize their emblematic role in depictions of the West African crisis. Journalist Robert Kaplan pictures migrants as voting with their feet, as creating a new world in the city amid disorder. "To these empowered millions, national borders, nations themselves, even the idea of a nation, were vague," Kaplan writes. "To them, the real borders were the most tangible and intractable ones—those of culture." Kaplan characterizes West Africa not only as a site of concern but also as a glimpse into the future, of the coming anarchy. Stark and sobering characterizations of West Africa depict the region as the most dysfunctional region in the world, as "*the* symbol of worldwide demographic, environmental, and societal stress, in which criminal anarchy emerges as the real strategic danger" (Kaplan 1994, p. 46).

Hyperbole notwithstanding, the concern for the stability of West Africa and the role of mobility in undermining that stability deserve serious attention. The presumed interconnections between environmental changes and the widespread urban migration in West Africa beg for more substantive treatment than the rhetoric used to drive policies of disengagement. However tempting it is to fall into a Conradian funk over the extreme conditions observed in the shantytowns of Abidjan or Lagos, one must bear in mind that much of the industrialized regions in Europe went through similar urban conversions. As sectors of the economy, "urban" and "rural" are in fact much closer than normally thought. They are not separate worlds, and it is not uncommon for a peasant worker to pass from one to the other and back on a regular basis. The social transformation on the individual level is often muted.

The future in West Africa is less likely to be a nightmare of refugees washing up on the shores of Europe than a slow and perhaps painful but probably silent process of social change. To shed light on the role of individuals within this change, and perhaps to find some lessons in their experiences, we must examine the material conditions of their worlds as well as their spatial mobility. What were economic and environmental conditions like prior to moving? To what extent were "uprooted" peasants already in motion? When Hausa peasants set off from their villages in November, where exactly do they go? Where do they sleep? What and where do they eat? Where do they work? When do they return? This study of the migrants of Maradi will sometimes ground, sometimes counter, and certainly humanize the futuristic and polemical visions produced by Kaplan and others.

Examination of the worlds of poor people in Africa and elsewhere reveals the invisibility and temporariness of their places, their lack of history other than family or lineage history, and their reliance on physical movement to alleviate stress and to obtain the resources to satisfy basic needs. The seasonal movements of people normally go unremarked by policymakers, nor do these movements appear on maps. It is common for people to travel long distances, especially to towns and to work locations by the roadsides, to seek work or alternative sources of income during the off-season when they are unable to grow food. These strategies are followed by the poor rather than the rich (De Waal 1988). Migration is usually considered a last resort, but in fact people often move into town when no work is available in the village and return afterward. The routine nature of this mobility activity is what renders it invisible.

The historical antecedents of today's observed mobilities and their association with poverty are difficult to investigate with existing historical accounts. The lack of detailed sources reinforces the common myth that prior to the modern industrial age the peasantry was immobile (Skeldon 1990). The lives of the poor are documented in historical records for Africa (Iliffe 1987), and in these accounts of itinerant beggars, hawkers, and social outcasts, the connection between poverty and mobility is made clear. The need for so many people to stay in motion is a dimension of human existence that is too often omitted from history. Moving people establish priorities that transcend fixed spatial coordinates.

Existing economic and demographic models of migration can identify why people decide to move, but they cannot explain the reasons for mobility itself. Accounts of forced population movements in history normally assume an ambient *prior* level of sedentarism. The victims of famine are normally assumed to have been rooted in place prior to the events that uprooted them. Particularly in non-Western settings, people use forms of mobility—especially circular ones—as de facto components of production systems. Urbanization and modernity have altered these forms, but the forms continue, and the relationship between changes in mobility practices over time and environmental changes, particularly changes in land uses and settlement patterns, begs for a more comprehensive treatment.

For a number of reasons, mobility is significant in understanding how people and places interact to affect food security, particularly where population and environmental stress are growing the fastest. Mobility challenges how residence itself is conceptualized, with settlements composed of sedentary inhabitants. The prevalence of circular migration in the African Sahel challenges the idea of a fixed population because the people enumerated in a village head count may not actually live there.

Mobility complicates the notion that a given area of land's "carrying capacity" is a fixed number of people who can be supported there. The notion that people migrate when carrying capacity is exceeded imposes a single measure of habitability and ignores the specifics of culture, economy, and the environment. For both conceptual and empirical reasons, there is no way to rely on a set number, or indeed any practical use for such a number once it is derived. Carrying capacity is difficult to quantify because absent people are unavailable. Those left behind may be unwilling to discuss their whereabouts with government authorities for fear of harassment or persecution. Often the absent people are counted anyway in censuses and demographic surveys. More to the point from a food security perspective, absent people may make demands on land by farming when they return, whereas present people may not, especially if they purchase food instead of growing it. Regarding an exceeded carrying capacity as the cause of involuntary migration ignores both the strategies migrants use to cope as well as the changes in areas affected by displaced populations.

Mobility also has potentially complex and significant relations to land use, with land-use change acting as both a cause and a consequence of mobility. Migration in Africa is a well-known response to income opportunities, a response that affects land-use and land-management decisions. The substantive question of mobility's consequences for agriculture has not received adequate empirical treatment. Neither has the contention that changes in land tenure and resource use associated with agricultural intensification displace people, only furthering the process of intensification.

The relationship between mobility and resource management shows that levels of mobility can perhaps be seen as a consequence of environmental stress or, conversely, that environmental stress or recovery can be seen as a consequence of mobility. Though sometimes regarded as a sign of degradation, mobility is also simply a traditional economic strategy. In areas with highly seasonal precipitation patterns, circular mobility appears to be a normal behavior (Hugo 1981). In Famine Early Warning System (FEWS) bulletins for the Sahel, observations appear from time to time about "hunger mobility" and the out-migrations of families, along with distress sales of animals and precious belongings and the eating of wild foods. Able-bodied men typically migrate seasonally to southern Zinder Department, Maradi Department, and Nigeria, but the out-migration of whole families indicates they have no assets left to sell or barter and that they have been forced to migrate (FEWS Bulletin, January 27, 1997).

The African Sahel has been portrayed as a portent of a global environmental future, and it is an oft-cited example of environmental ruin. Prior to the devastating drought that occurred there in the late 1960s, the Sahel region was virtually unknown to the Western world. While the conditions there were capturing the world's attention and begging for its compassion, an entity called "the Sahel" was created, first by French anthropologists and international aid workers. French structuralists Claude Meillassoux (1974) and Jean Copans (1983) looked behind the faces of famine victims to find those who profited from the exploitation of natural disasters. Copans writes, "We shall understand the true effects of drought when we can show to what extent local bureaucracies and bourgeoisies have utilized them in order to consolidate their power. What is called humanitarian aid is in fact an astute (and cheap) means of preserving domination and a beggar's mentality among the dominated" (1983, p. 94). Comments like these influenced researchers like Kenneth Hewitt (1983) and Michael Watts (1987), who attributed the cause of disaster to capitalism and developmentalist greed.

What visitors witnessed during those times of drought were massive out-migrations of entire villages and clans, the decimation of livestock herds, and widespread human suffering. What the visitors did not see occurred later, as farmers and herders returned to their homes after the rains had begun again. The resilience of the people was not in evidence, and the relatively low loss of life from the multiyear drought likewise escaped the attention of most observers and commentators.[5] This resilience, which is bred into the backbone of the cultures of the Sahel, has permitted the continued settlement of a region inhospitable to many activities.

My own relationship with Niger began in 1985, about ten years after the 1968–1974 drought and one year after the 1982–1984 drought devastated the region. I lived for two years in Guidan Roumdji, a large town about fifty kilometers west of Maradi, and I taught English as a Peace Corps volunteer in a middle school there. In my twenty-two-year-old naiveté I imagined Niger as a neutral zone, a place where I could take refuge for a while in anonymity. My first impressions, which I recorded dutifully in my journal, were of a barren land, something that I perversely imagined to be postapocalyptic. Early on I noted the intriguing way in which distance in the vast (almost twice the size of Texas) and sparsely populated land was used. People in Niger are accustomed to living far apart, but children are also comfortable being scrunched together on a bench to watch a wrestling match. I noted that car parts were used as livestock fencing and that the kilometer markers on the national highway are shaped like tombstones.

It took me a while to get to know the people. I had severe language difficulties: My high school French was hardly adequate and learning Hausa, the really useful language, took some time. When I did become acclimated, the place grew on me and took me over. Then, seven years after I said good-bye to Niger, I left my own family behind and returned to conduct research on the people who make their living by migrating seasonally—not just wandering herders, but the supposedly sedentary farmers who actively engage in the market economy. I undertook four months of data collection and interviews in the fall of 1995, then I returned again for more fieldwork the following spring. I was determined to find a world beneath the surface—beyond the gaze of demographic surveys and satellite photos—a social world constructed around elaborate networks based on family ties, friendships, and trust. My aim was not to romanticize the lives of the poor but simply to detail them and to be realistic about them.

Ample baseline geographical research had been done in Maradi by a cohort of French researchers, led by Immanuel Grégoire and Claude Raynaut, from the University of Bordeaux. From the early 1970s to the mid-1980s, these geographers and social scientists operated a post in Maradi and conducted many studies (Koechlin, Raynaut, and Stigliano 1980; Grégoire and Raynaut 1980; Brasset, Koechlin, and Raynaut 1984 Raynaut 1988; GRID 1990) with the assistance of the provincial government. In the departmental planning office, I discovered studies of land use patterns for the Maradi study region that proved to be very useful for my own research. These archival data provided a baseline for changes in land use and intensity over a twenty-year span (1975 to 1995), which helped document the loss of fallow and bush land. My research affiliation with the Agro-Meteorological and Operational Hydrology Center (AGRHYMET) in Niamey allowed me to analyze aerial videographed flyovers covering specific areas in Mayahi *arrondissement* (district), where I had done the village interviews and whence many circular migrants came. By adding the videographed data to digitized archival data in a geographic information system (GIS), changes in land-use intensity between 1975 and 1995 could be compared with a high level of precision. This could then be used to determine where land uses had changed or fallowing had been abandoned because of population pressure.

Data on the changing political-economic context of circular migration in Maradi Department were obtained from a number of sources in Niger. I interviewed the then-civilian *préfet* of Maradi and the *sous-préfet* of Mayahi, personnel in the mayor's office in Maradi and the state-run transportation service, and many other informants in Maradi and Niamey. In the development community, I questioned representatives of the U.S. Agency for International Development (USAID), the Famine Early Warning System, the Disaster Preparedness and Mitigation Program, the International Food Policy Research Institute, and the U.S. Peace Corps. I relied on published and unpublished documents for statements about development policy in Niger, as well as documents from the Projet Maradi team and existing literature on Hausa economy and culture.

To investigate the changing worlds of circular migrants in Maradi, I developed profiles for the individuals in Niger who migrate to nearby cities in the dry season. In November and December of 1995, I interviewed 133 labor migrants in the city of Maradi. With the help of the Maradi mayor's office, I first identified certain later-settled neighborhoods of Maradi, sometimes on the outskirts of town, that had initially accommodated the large influxes of migrants after the droughts of 1968–1974 and 1982–1984 (see Photo 1.1). I hired a former census-enu-

Photo 1.1   Migrants' housing in Maradi in southern Niger. Circular migrants often find a place to live inside existing compounds and will sometimes work sporadically for the homeowner.

merator, Kanta Wakasso, to locate potential sites around the city for interviewing circular migrants. After asking their permission and obtaining their informed consent, Kanta assisted with the interviews and translated responses from Hausa into French. The forms contained questions on the migrant's life in the city, work types, income levels, expenses, family information, contact with people from the village, gifts and investments in the village, and migration routes. Many of the questions were open-ended. Migrants' life histories and stories were recorded in written notes during and after the interviews. I made deliberate attempts to contact and interview as many women as possible, but even so the sample group, as well as the overall migrant population that it can be thought to represent, is around 90 percent male, consisting mostly of Hausa-speaking agriculturalists from the drier northern districts within the Department of Maradi.

My interest in the particular relationship between migration and food security in the village was sparked by many conversations with the migrants about their lives and their movements. The answers elicited from questions about intentions seemed inadequate to describe the decision-making process behind the movements, and this prompted further investigation in both the city and some of the sending villages. When I returned to Maradi again six months later, I sought out some of the migrants I had interviewed the previous fall to ask follow-up questions about the money they remitted and about changes in their employment. One of my aims was to estimate the consequences of the increased amount of labor that circulates between the city and villages of Maradi Department and the uses to which resource flows associated with this circular mobility are put, especially on village agricultural production. With this I could determine the extent to which migration has become a necessary component of individuals' and communities' survival strategies and the effect this is having on food security. To prepare for the village work, I constructed land-use maps from the 1955 and 1975 aerial photos to document changes in land cover. A second objective was to probe the use of social networks to combat the risk of periodic drought and famine. This final stage of the fieldwork included a series of stays in the village of Guidan Wari in Mayahi district.

Geography Lesson

Niger is a landlocked country lying between the Sahara Desert and the tropical savanna in West Africa. Only a small portion of Niger's total land area is arable, with 2.85 percent in permanent crops, 8.24 percent in pasture, and 2.0 percent in forest and woodland (FAO 1995). The remainder is unused desert wasteland. More heavily populated regions are marked by increasing land scarcity and population pressure (Raynaut 1988). A survey of villagers in the Baban Rafi forest in southern Niger found that a significant number did not expect their forest to outlast their lifetimes (Elbow 1992).

Like most countries in the Sahelo-Sudanian agricultural zone, the food system of Niger is based on nonirrigated agriculture using hoes (see the Sahel rainfall zones map in Figure 1.2). Annual precipitation is determined by the northward extent of the monsoonal intertropical convergence zone, with rainfall confined almost entirely to two to four months during the summer. Since the mid–1960s, the Sahel region has experienced annual rainfall between 20 and 40 percent less than between 1931 and 1960, although the 1994 rains were double the normal amount.

The portion of Niger that lies in the Sahelian zone, north of the fourteenth parallel, was not settled until the start of the present century, owing to security considerations as well as to unreliable conditions for farming. Settlement occurred after the French pacification of the region's pastoralist Tuareg population and the establishment of a rudimentary communications and transportation infrastructure. The settlement of the Sahelian zone occurred with the indirect complicity of the climate. Encouraged by

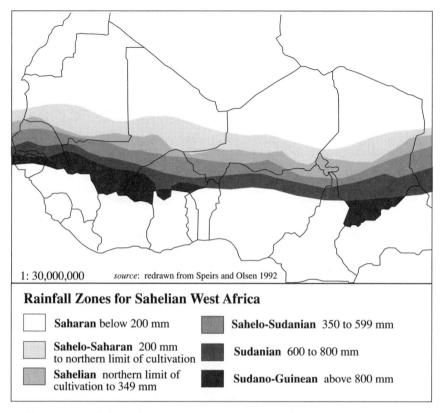

FIGURE 1.2   *Sahel Rainfall Zones*

higher-than-average rainfall from the 1930s through the early 1960s, sedentary Hausa farmers expanded into drier northern zones and kept pace with expanding food needs by exploiting margins of new land (Delehanty 1988). About the same time, Tuareg and Bougajé pastoralists descended from the north and settled. Rural areas were characterized by shifting agriculture, with family groups moving about on the open-access land in search of game and undisturbed soil, settling for a few years, and then leaving as soon as the land became less fertile and the hunt less productive (Raynaut 1988). This system of bush fallow allowed margins of land to be exploited when rainfall was better than average, and it also allowed ample room for livestock and trees. This land-use system lasted roughly until the early 1960s, when the frontier had effectively closed to new village formation. As villages subdivided in order to cultivate bush land lying between settlements, open space was turned over to crop agriculture and gradually became almost completely filled.

Although the population of Maradi Department more than tripled from 1959 to 1996, the production of four main crops—millet, sorghum, cowpeas, and groundnuts (peanuts)—kept pace with the increase, growing three and a half times in the same period, mainly through expansion of cropland rather than increased yields. This growth occurred throughout most of Maradi, with the areal extent of farming increasing sevenfold from 1959 to 1995, from 387,290 hectares to 2.7 million hectares (Direction de l'Agriculture 1996).

The spread of settlements and natural increases in population resulted in the creation of land scarcity in areas where once only people and water had been scarce and land had been abundant. As uncultivated bush land became increasingly limited, peasants discovered that they could not manage to fill granaries, market surpluses, and pasture livestock at the same time without a margin of free land to exploit periodically. This reduced their capacity to adapt to climatic fluctuations and created a crisis of chronic shortfalls, which was compounded by several episodes of drought, particularly from the late 1960s to the early 1980s.

Worsening the erosion of rural food sufficiency was the growing economic reach and food demands of the departmental hub of Maradi, which gradually incorporated the countryside into a larger economic system based on kinship, affinal, and market ties. In the midnineteenth century, the city of Maradi was nothing more than an isolated community of a few thousand *Habé* (animists) living in the shaded refuge of Goulbin Maradi, a seasonal river valley. These refugees had been banished from the seven Hausa states of the Sokoto Caliphate (the "Hausa *bokwai*"), in what is now Nigeria, by a terror campaign and religious war incited by the Muslim leader Usman dan Fodio[6] in the early years of the nineteenth century (David 1969; Thom 1975).

French colonial control of Maradi came in the early 1920s with the opening of a small garrison and living quarters for European military officers (Grégoire 1992). The military presence allowed for the installation of European commercial houses after 1924, at which time groundnut seed was distributed. By 1927, a commercial network was established, and eleven trading firms were in business by 1931. The road and communications networks, as well as the new cash crop, were restructuring the economy of the south-central region, and the groundnut boom of the 1920s ensured the growing commercial importance of the city given its proximity to the British territory of Nigeria (Grégoire and Raynaut 1980).

Raynaut (1988) sees colonial interventions as being critical to the transformation of Maradi Department and of northern Hausaland in general. The colonial interlude was driven more by geopolitical imperatives— namely, the rivalry between the British and French—than by the attraction of natural resources. In fact, the only true resource mentioned in

Maurice Abadie's early history of the French military territory of Niger was the human inhabitants of the territory themselves and their labor power (Abadie 1927). Raynaut identifies the two imperatives that guided development policy for the territory since the start of European colonization: first, to open peasant society to the market economy; second, to allow the population growth rates to increase by reducing mortality. The golden rule of development in the Niger *territoire militaire* was to cultivate the labor supply and then to give it an incentive to leave (Raynaut 1988). Looking back on Abadie's early account, it is striking how correct the colonial blueprint ironically has been. In the history of Niger thus far, the only exploitable resources other than human labor and ingenuity have been livestock, grazing lands, and uranium. Uranium enjoyed a short-lived boom in the northern part of the country, which lasted from the mid-1970s until the Three Mile Island disaster reduced the worldwide demand for the radioactive fuel.

The relationship between the city of Maradi and the country of Nigeria, its neighbor to the south, has been close in colonial and postcolonial times. Maradi became the hub of a regional economy based in part on black market trade in gasoline, cigarettes, and plastic housewares, as well as food from the villages of Maradi Department (see Photo 1.2). Nigerians' investment in the industries and products of the city of Maradi increased especially during the Biafran war, when trade to areas of conflict in the Niger River delta was interrupted and interest turned north (Grégoire 1992). Monday and Friday market days in Maradi are notable for the presence of large numbers of Nigerians, who come to town to purchase, with Nigerian currency, such products as meat, animal skins, cowpeas, and millet, which they stow in the trunks of their Mercedes-Benz automobiles. Though by no means on an equal footing, Maradi and Nigeria have socioeconomic ties that run deep, and market activity explains in large part the current vitality of the city of Maradi.

Maradi's course changed with the droughts that began in 1968 throughout the department and across the entire Sahel. Villagers trying to escape increasingly desperate conditions during the drought years begged on city streets and found work in the informal economy. When one period of drought ended in the early 1970s, the city had grown to 44,458. By 1981, it had grown to a permanent population of 66,472 (Ministère du Plan 1993). The revenues from uranium mining provided the money to expand the city substantially, allowing for the construction of new schools, an improved market, and a sewer system. With its continued black market dynamism, the departmental capital of Maradi has come to dominate ever more territory in south-central Niger.

By observing the role played by circular migrants in the process of changing urban-rural relations, we also see the handprint of West African

Photo 1.2   The Nigerian trucks on a Maradi street near the market exemplify the role played by the economy of Nigeria, the elephant to the south, in the daily existence of the border town.

urbanization. In Niger, the proportion of the total population living in cities and towns rose from 1.3 percent in 1930 to 5.3 percent in 1960 to 15.9 percent in 1990, and it is expected to rise to 30 percent by the year 2020. Within Niger, urban growth ranges from 4 percent to over 10 percent annually. The extent of world urbanization is usually represented in the growth of "millionaire-cities," or megacities, but West Africa's ongoing conversion to urban life is better seen on a smaller scale, with the number of cities with 100,000 inhabitants or more having risen from five in 1930 to seventeen in 1960 to sixty in 1990 and being forecast to reach three hundred in the year 2020 (WALTPS 1994).

Urbanization continues amid increasing political and economic marginalization for sub-Saharan African countries in general, particularly for the West and Central African countries dependent on France. The fourteen countries that make up the Communauté Financière Africaine (CFA) monetary zone found their futures destabilized at least temporarily with the devaluation of the CFA franc (FCFA), an African currency tied directly to the French franc, in January 1994. This occurred shortly after the death of Felix Houphouët-Boigny of Côte d'Ivoire in December 1993, the staunchest African supporter of close Franco-African ties. It also marked the ascendance of the International Monetary Fund (IMF) in what was formerly considered France's *domaine réservé* (Schraeder 1997). Further deterioration in terms of trade, the alienation of former French colonies from patronage (including business concessions and scholarships for African students), and the uncertain future of the FCFA given the monetary strictures of France's membership in the European Union, have increasingly caused the francophone African states to appear to be cast adrift from the rest of the world.

Internally, factors such as development policies that squeeze small producers through the loss of aid programs and subsidies, the failure of democracy campaigns since their promising early start in 1990 and 1991, and the return to personal rule by autocrats (Charlick 1991) signify that migration has become a necessary means of political expression. Some people have been left with no recourse but to pick up and leave. The brain drain is especially acute, since educated professionals and business people take advantage of opportunities elsewhere. The economic and political woes endemic in West Africa have closed opportunities for less-favored Sahelians in the coastal cities as well, prompting many to migrate to cities within their own countries.

At the same time, however high the current rates of urban growth are, urbanization is not new to the region of southern Niger where this study takes place. In a treatment of traditional Hausa urbanism, Ahmed Beitallah Yusuf (1974) identifies seventeen major types of settlements along a continuum ranging from *dokar daji* (uninhabited wasteland) to *birni* (city-state). Urbanism has been a feature of Hausa culture since the great trade terminus of Kano, in northern Nigeria, was founded over a thousand years ago. The European explorer Heinrich Barth estimated the population of Katsina in 1851 at over 100,000. The special features of Hausa urban life, such as cultural heterogeneity and acceptance of social difference within the city, elaborate systems of tribute and taxation, and the pervasive enabling mechanism of Koranic education, are well known.

The effects of urbanization are most observable in the coastal cities of West Africa, but the changes fueling these effects are by no means confined to these areas. Maradi is marginal to the world system. It produces few goods for world trade, and unlike other African cities such as Accra or Lagos, Maradi does not react in a frenzied fashion to the rise and fall of international commodity prices. Yet within that economically marginal predicament, it is experiencing profound and varied changes in production, social relationships, and people-land relationships. Some towns and villages in the surrounding region have grown 10 percent per year for a decade or more, whereas others have lost numbers of people. Some people have experienced increased food security and well-being, whereas others have become increasingly hungry.

The Focus on Individuals

One empirical challenge of this study is to relate the changes in individuals' worlds to changes in environments and economics at the local and regional levels. The strategy is to use data collected at multiple levels contextually to relate the individual perspectives on circular mobility to regional change. Social networks, because they are difficult to measure, and individual biographical details, because they often appear random, can hardly be thought to fit into a framework of regional change. Yet a focus on individuals is justified because individuals are obviously the most careful monitors of their own situations, and situational knowledge is most definitely affected by regional dynamics.

Understanding the interlinked relationship between population mobility and settlement-level resource use cannot occur without a comprehensive treatment of people—including their beliefs and actions—and of how their social power operates across space. Contextualizing people within the environments they inhabit will reveal how environmental change favors some individuals and groups over others. Such individual content can help explain why, for instance, people will migrate with the intention of finding food even though food is readily available at the local market. A perceptual gap exists between macrolevel regional assessments and microlevel popular responses, and this gap has a scale and a geography of its own. As the following profiles show, people do not naturally fit into a social science model. Like most people, migrants do not see the "big picture" of environmental influence. This does not mean that elements of the big picture do not motivate them; rather, it means that they will perceive these elements only as filtered through their own experiences. Migrants do not, for instance, relate the sequence of settlement of their villages of origin to the colonial construction of roads and other infrastructure in the area decades before. They do not relate their movements to the erosion of village-level food sufficiency, nor do they associate an increasingly precarious existence with the actions of their national government or development agencies.

Some examples from the city of Maradi illustrate this point. I talked with Malam Harouna Issa,[7] a Hausa-speaking *marabout* (Islamic priest), in the shade of a millet-stalk shelter in an established neighborhood, with a group of perhaps ten men and boys crowded around to listen. Malam Harouna was born in 1941, and he left his village of Chadakori in 1972 after several years of below-average rainfall. With the family granaries empty, his wife and their children accompanied him on foot to the city, about fifteen kilometers away. After arriving in Maradi, Malam Harouna begged for his food until he found people in Maradaoua *quartier* (neighborhood) willing to support him. Now a religious leader for his neighborhood, he trains local children in Koranic teachings and the Arabic alphabet. He states that he has little contact with his village of origin except for the occasional gift of millet from his extended family there. In return, he sends his family in the village clothing and occasional gifts.

Insufficient rainfall was clearly the trigger that motivated Malam Harouna's initial move to Maradi. He and his family sold their assets, ate their food reserves, and exhausted the possibilities of the social safety net in Chadakori before deciding to travel to the city. But however clear-cut the decision he and his family made to leave the village, Malam Harouna would not dwell on the specific conditions in his village of origin that made him decide to remain in the city and not return to Chadakori. Although he has now spent the better part of his adult life in Maradi, Malam Harouna steadfastly refuses to speculate about whether his stay is permanent or not. Such feelings about the impermanence of a move are very common among the migrants I interviewed. Many did not acknowledge that they have finally left their village homes. In a way, many haven't.

Hawa Hada is a female medicine seller, aged forty-two. Hawa is Bororo, or bush Fulani, born in a hamlet north of Dakoro in the far north of Maradi Department. She began migrating seasonally in 1973, passing through Maradi to the West African coast, during the same drought period as the one that motivated Malam Harouna to quit Chadakori. She is married and has two older children, but she left her family and their flock of animals behind to travel. She has gone on seasonal migration for the last twenty-two years and has visited cities in Cameroon, Benin, Nigeria, and Côte d'Ivoire, traveling every year to plait hair and sell traditional herbal medicine. When I met her, she had arrived in Maradi just the day before. She had plans to leave the following day for Benin, traveling in a group of twenty-five other Bororo women. Her liking for travel was so strong that she said she would still go away even if there were paying work back home. While drought has influenced Hawa Hada's decisions from time to time, she maintains that she would migrate anyway.

With Hawa Hada, the environment acted as an initial trigger but not a shaper of mobility decisions. As a pastoralist, she inherits a family history and holds a worldview that depend on the ability to change locations to take advantage of spatially separated resources and opportunities. As a Bororo woman, she enjoys freedoms not usually available to Hausa women: She has far more autonomy, a more egalitarian relationship with her husband, and fewer children than women from traditionally grain-growing cultures. Her case highlights the importance of ingenuity as a resource. By improvising, Hawa Hada has made the most of what seems like a difficult situation; this is a common feature of Sahelian circular migrants, and perhaps of migrants in general. Her case also high-

Photo 1.3   *Mohammed Keita
stands in front of the products of
his business, metal piping and
reinforcing bars, on a major
route near the center of Maradi.*

lights the role played by cultural norms in affecting the prevalence of circular mobility.

My third example is a merchant who successfully straddles urban and rural life. Mohammed Keita (see Photo 1.3), though still in his late twenties, is a man of wealth and influence. Leaving his village initially to go to school in Maradi and formally educated to the high school level, Mohammed left school several years ago after a succession of incomplete school years (*années blanches*) caused by a prolonged government crisis and the stoppage of teacher pay. With the help of his social contacts, he began work as a trader. Mohammed now owns a commercial metal business that supplies reinforcing bar and piping to Maradi's thriving construction sector. He is from one of the established families of Guidan Wari, in the Sahelian agro-ecological zone located ninety-five kilometers to the north of Maradi. Though residing in Maradi year-round, he does return to his village for social occasions. He does not farm there anymore. Like many in Maradi's capitalist class, he engages in speculative buying and selling of grain and other commodities in the city. Mohammed watches the market shrewdly and gets information about prices through his network of friends and acquaintances

Mohammed is central to his social network, occupying a pivotal point of contact for virtually any resident of Guidan Wari who comes into Maradi and serving as a source of news from the village for year-round Maradi residents. Mohammed does not now travel much himself, but he does benefit from the movement of others. He has adapted exceptionally well to an uncertain physical and economic environment by diversifying his investments and maintaining a large network of contacts, allowing him to spread his interests and risks. Speculating successfully requires luck and good sources of information. Ironically, his financial interests do not extend to agricultural production in his village of origin because he feels the risk of failure is too high. So, although the environment of Guidan Wari strictly speaking did not influence Mohammed's decision to migrate, it does affect his reluctance to invest in his village of origin.

These sketches show the variety of Maradi migrants' experiences. If any common feature binds them, it is that they swim in a sea of affinal, kinship, and economic network contacts. With this in mind, concentrating on the migrants only in the context of their families or their regional environments is perhaps not sufficient. To understand the migrants truly, one must set the characteristics of individual movements into their larger contexts by looking closely and specifically at the nature of social ties. As this study will show, social ties are often far more immediate, and perhaps more unpredictable, than environmental conditions alone in determining the success or failure of actions.

Of course, migrants may have other motivations. Fights with one's father, marital infidelity, disputes over inheritances, the failure to be chosen as new chief, the death of one's children from measles, or other misfortunes are often what motivate people to go away from the villages they call home. They are also often motives that the migrants are understandably reluctant to discuss.

The complexity of mobility is tied not only to the distance and duration of the moves but also to what the migrating individual considers "home" to be. Notions of residence are difficult to establish when individuals are permanently in motion. The nature of the migrants' activities contributes some complexity as well. It should be obvious that in typical low-income settings, many do not commute to work in the way that those in industrial countries do.

Underlying the complexity of mobility there are distinct geographical implications, relating not only to notions of residence and livelihood but also to how places are defined and perceived and how social ties within these places are valued. Our awareness of the critical role of social networks will be called upon when we look at the particulars of mobility in sub-Saharan Africa, where the problem of multiple destinations and multiple affiliations is especially critical. At the level of individual people and places, the roles of social networks, nature, and culture are too important to be dismissed.

Mobility as Archetype

People eat and people move, but we cannot talk about their doing so without being almost smothered by theory that fails to recognize the nuances of their actions. Regarding the subjects without ideological blinders is evidently very difficult, since few have resisted them. The theories concocted tend to have the effect of promoting one ideology over all others, usually to the detriment of an understanding of the migrants themselves.

The political-economic critique of development theory and of subsequent explanations for migration comes from the "dependency" school, formed in the 1960s by theorists studying Latin America (Frank 1967; Dos Santos 1971) as a reaction against the then-reigning "stages of development" school (Rostow 1952), which portrayed human societies around the world as progressing through a predetermined set of evolutionary levels and culminating in a civilization resembling the West. The dependency school explicitly acknowledged the role played by the First World—the industrialized nations of the West—in restructuring the developing nations of the Third World, making them dependent on the West. In his book *Neo-Colonialism in West Africa* (1974b), the dependency theorist Samir Amin based his explanations for changes in West Africa, including changes in migration patterns, on the geographical location of "factors of production," including labor, capital, natural resources, and land. These factors should not be accepted as existing a priori, Amin argues; rather, they are the result of a deliberate strategy of development pursued by European colonial powers.

Amin's argument is that African economies collided with colonial capitalism and were altered in striking ways, with profound consequences for mobility. The special nature of colonial capitalism on the economic periphery makes the forms of mobility there particularly exploitative. Circular migration permits capital to circumvent paying indirect wages, which are costs related to reproduction of the labor force and its maintenance in the early and late years of life. Instead, capital must pay only direct wages, which are those necessary only for daily reproduction of labor power. In peripheral capitalism, indirect wages are borne entirely by the domestic sphere (Meillassoux 1981b; Cordell, Gregory, and Piché 1996).

By this strategy of labor exploitation, the modern map of West Africa was redrawn as a multisectoral land-use plan conceived by European capitalists. Coastal West Africa became dominated by resource extraction and plantation economies oriented to European export, and the interior functioned as a massive labor reserve. Interactions between the coast and the interior were built on centuries-old commercial ties based originally on slave trading. The rise of port cities and transportation corridors, including roads and railroad lines, permanently altered countries' political geography. Head taxes, forced labor, the conversion of local economies to the use of cash, changes in agricultural systems and land-use patterns, the restructuring of household economies, rudimentary improvements in public health, and changes in gender relations and notions of property, customs, and power are all consequences of the colonial era.

In Amin's dependency explanation of population mobility in West Africa, modern population movements are the result of regional inequality and uneven development instigated by colonial capitalism. In *Modern Migrations in Western Africa* (1974a), Amin distinguishes four kinds of population mobility in West Africa: first, coastal urban movement, which is dominated by labor migration; second, the traditional movements related to the colonization of new lands; third, population movements associated with the "Nigerian exception," which is distinctive for its high levels of north-south circulation; and fourth, what Amin calls "exodus," which spans the categories and includes wandering herders and persons displaced by drought or political doings.

In his quest to present migrants as victims of colonialism, Amin downplays the consequences of circular mobility in a typical way. He dismisses rural-rural circulation as not "real" migration because seasonal migrants who go away in September and return in April or May often have two homes, so they cause little demographic impact. Amin likewise dismisses mobility undertaken for commerce as "precolonial" and hence not worthy of serious consideration. Amin writes that modern migrations in West Africa involve "labour, not people" (Amin 1974a, p. 66), by which he apparently means that migrants come into a receiving society that is already organized and structured. "Labour" is one step removed from the analytical scale of "people," and in that step individual ingenuity and strategies are likely to be lost. One cannot help but feel that by turning migrants into the abstraction that is "labour" Amin deprives them of some of their humanity.

Amin's dependency thesis wields explanatory power for changes in migration patterns when fleshed out with historical and geographical particulars. Immanuel Wallerstein's world-system theory can be viewed as an improvement upon dependency theory that puts more emphasis on transformations over large spaces and over the *longue durée* (extended course) of history (Wallerstein 1974; Yo 1990). Under the theoretical organization of dependency theory, or its revision in the form of world-system theory, mobility is conceptualized as a consequence of economic restructuring.

For his study of Zarma peasants from the Department of Dosso in southwestern Niger, development sociologist Thomas Painter utilizes world-system theory to illustrate the transformation of the Sahelian peasantry (Painter 1985, 1987). Since the turn of the century, peasants have been drawn into the world capitalist system through seasonal migration to Côte d'Ivoire, Ghana, Nigeria, Benin, Togo, and Cameroon, impelled by corvée and head taxes, which dramatically increased the need for cash in households. Painter's methodology for measuring the modern implications of reliance on migration is to determine the proportion of household income from off-farm sources that was spent on subsistence grains. Painter finds this proportion to be roughly 50 to 80 percent. Painter concludes that leaving the land in the off-season did little to alleviate rural poverty and instead perpetuated seasonal vulnerability to food shortages by diverting energy that could be invested in field preparation, land improvement, and village capital formation. Painter devotes a section of his contemporary analysis to official statements from the government of Niger calling for the abolition of seasonal migration, which has been called a "sickness" and a problem with "important and alarming" consequences (Sidikou 1978, quoted in Painter 1985). Painter captures the mood of the 1970s, when seasonal mobility was publicly denounced as a colonial remnant and peasants were told they belonged in their villages. Policy in Niger to a certain extent continues to reflect this view of mobility. Both the government of Niger and international development agencies maintain their conviction that greater agricultural productivity and natural resource management *could* be achieved if people could be induced to remain in their villages in the off-season and to migrate less (Taylor-Powell 1992).

However compelling, the dependency explanation for population mobility ignores the fact that mobility predates the era of colonial capitalism in the region by centuries. Painter dismisses the historical continuity argument of Jean Rouch, who has documented the movements of small groups of Zarma warriors in and out of Niger from at least the 1860s (Rouch 1950, 1956, 1961). Rouch researched the long-distance movements of Zarma men employed as mercenaries to Dagomba in the northern Gold Coast. As one result of these movements, links were established between what is now southeastern Niger and Dagomba, Salaga, and areas further south in the Ashanti kingdom (Olivier de Sardan 1984). Rouch argues that these precolonial links made possible subsequent migrations from southwestern Niger.

Dependency theorists de-emphasize the fact that seasonal migrations had been prompted by the needs of seasonal agriculture long before the colonial period (Coquery-Vidrovitch 1988) and that a substantial trade diaspora existed prior to European conquest (Bohannon and Dalton 1962; Meillassoux 1971; Hopkins 1973; Lovejoy and Baier 1975; Lovejoy 1980). Painter argues that the importance of established networks for understanding modern migrations is exaggerated, and by doing so he effectively rejects historical continuity out of hand.

Assessing dependency theory as it pertains to today's changing mobilities is difficult, for one might agree with the conclusions even as one contests the analysis. In recent years, the theoretical literature has become more subtle, and whether one sees the consequences of migration as positive or negative depends on one's perspective (Cordell, Gregory, and Piché 1996). Nevertheless, the conclusions reached by Amin and Painter are compelling, partly because there is a smoking gun of sorts in the bald European colonial designs. There is also a coherent explanation for the vague sense of something being wrong in countries in the Sahelian periphery. Even so, dependency explanations for mobility, and structural explanations in general, tend to follow a single chain of exploitation paradigm (Kearney 1986), which entails a one-dimensional replacement of traditional modes by capitalist modes.

Dependency theory, if applied to specific cases, highlights what is still a deep-rooted influence of the colonial era upon the population geography of the West African region. Given independence without the civil institutions to establish a cohesive national identity, governments in many West African countries lack the authority or ability to inspire people to follow willingly. Porous country boundaries do little to stop the flows of people and goods across often artificial colonial frontiers. Social networks are omnipresent, and they sometimes actively subvert the actions of the state. The concept of residence is subject to differing interpretations and is based on social ties that are subtle, complex, and changing. Each of these regional features is constituted by, and is reflected in, the particular ways people move about to satisfy needs.

If dependency theory tends to neglect the role of historical continuity in determining modern migration flows, a similar charge can be made against neoclassical theories of population mobility, which are configured around expectations of income gains from migrating. Michael Todaro's neoclassical premise in the microeconomic theory of migration is based on rational economic calculation, with migration decisions forming a response to rural-urban income differentials in expected earnings (Todaro 1969). Todaro calculated expected earnings of migrants as the product of the wage rate in the modern sector multiplied by the probability of getting a modern-sector job over time. The process of gaining access to the income justifies a time-specific approach. Immediately after arriving in the city, migrants have a remote chance of finding employment, but after remaining there for a period of months or years, they become acquainted with social networks and incrementally increase their chance of becoming employed. Jeffrey Williamson's critique finds Todaro's theory riddled with implausible assumptions (Williamson 1988). According to Williamson, Todaro's model simplifies the mechanism of job allocation and decisionmaking and ignores the informal labor market. For present purposes, it is the lack of understanding of the informal economy that is the most glaring omission.

The informal economy resists a single definition, but it is generally recognized to be any economic activity not recorded in national income accounts of most countries (J. Thomas 1992). More precisely, it is a set of economic activities often but not exclusively carried out in small firms or by the self-employed, activities that elude government requirements such as registration, taxes, social security obligations, and health and safety rules (Roberts 1989). According to the United Nations Development Program (UNDP), 60 percent of the labor force in Africa is employed in the informal sector, and this force generates 20 percent of the gross domestic product (GDP) (UNDP 1997).

Like Amin and Painter, Todaro argues that heightened levels of urban migration have negative consequences for rural sending areas. Elliot Berg's neoclassical explanation of the economics of the migrant labor system (Berg 1965) presents a different view of the migrant labor system from Todaro's. In Berg's view, although the West African migration system is often pictured as a major factor contributing to high labor turnover, low skill levels, low wages, and the backwardness of agriculture, it is an efficient adaptation to a particular context. With migrants available from many sources and with large sensitive labor markets ready to enfold them, Berg sees West Africa as a place where social and environmental costs, such as neglect of village capital formation and agricultural field preparation, are outweighed by gains in income and innovations brought back from migration destinations. Pleas for understanding the inherent rationality of seasonal migration echo his original view: "The prevalence and persistence of the migrant labor system, then, do not arise from the perversity or wickedness of men. The system has a secure foundation in the economic environment of West Africa. It cannot be wished away" (Berg 1965, p. 181).

The economic circumstances that Berg alludes to are not necessarily different from those postcolonial conditions highlighted by dependency theory. Berg's assessment of the migrant labor system is realistic, and it continues to be the sentiment of many economic development practitioners in Africa (Becker, Hamer, and Morrison 1994). According to Oded

Stark, rather than trying to devise more effective measures to contain or reverse the trends of mobility, development analysts ought to turn mobility into a vehicle of national development and personal betterment (Stark 1991). Explanations of exactly how migration makes good economic sense tends to neglect other issues besides the standard neoclassical emphasis on income differentials.

Neoclassical migration theory is difficult to test in cases where circular mobility is utilized to spread interests among multiple destinations and where ties to the origin are not cut. Neither does the theoretical approach account for changes in economic or environmental conditions that alter population flows, other than counting worsening rural conditions as "push factors." Basing mobility decisions on such stimulus-response calculi ignores individual and group strategies and renders deterministic the decisionmaking process. Orthodox migration approaches are most useful for isolating determinants in a cross-sectional sample designed to be representative nationally. Although they are often appropriate for the stated objectives, such approaches cannot fully explain the economic causes or consequences of migration in terms other than income gains or losses.

Stark's retooling of neoclassical migration models points out this shortcoming, explaining that although migration is considered an individual activity, the mechanisms are too complex to be understood simply as individualistic optimizing behavior (Stark 1991). Instead, the migrant must be seen to make decisions based on mutual interdependence, usually on intrafamily exchanges that are integral to mobility. Such exchanges are evidenced in remittances, destination preferences, and the various constraints put on family members. In the absence of wage differentials, migration does not necessarily mean irrational behavior; rather, it might reflect income uncertainty, relative deprivation, family pooling of risks, or human capital investments in children (Stark 1991).

It is no coincidence that circular mobility behavior often occurs in places where resources are spread far apart and families are large. Choosing migration in such circumstances is eminently rational and strategic, especially when human capital can be distributed across several markets. The theoretical challenge of linking mobility and the regions in which it occurs is made more pressing by questions about the consequences of mobility for the rural sector. Do the money, goods, and services sent back to the villages by migrants simply prop up production systems that are already overstressed, or can they assist in the long-term development of the village?

Jane Hopkins and Thomas Reardon have argued that income from seasonal migration is financing investments in agricultural production and sustainable change in the Sahel (Hopkins and Reardon 1993). Hopkins, an agricultural economist from the International Food Policy Research Institute, regards migration as a method to increase outside income flowing into rural villages. Migration remittances provide a kind of resource circularity in which the nonlabor counterstream from rural-urban migration—goods, cash, food, tools, materials, knowledge—balances the loss of labor from migration.

According to this conception, the village labor lost to migration is replaced by flows of other resources, particularly goods that enhance the long-term sustainability of agriculture. Sally Findlay's research in Mali found that some long-term migration forms favor capital accumulation, which for richer farmers can be used to buy tractors, improved plows, and irrigation pumps (Findlay 1987, 1992). These questions about the role of migration remittances in bolstering the economy of rural villages are critical and will be explored in detail later.

In West Africa, the spatial distribution of mobility systems, people, and resources occurs in an environment that is also heavily influenced by the climate. This is particularly true for Sahelian mobility systems. The temperature and precipitation regime of the dry interior of West Africa is characterized by an extremely seasonal and unimodal distribution of rainfall, which leaves a "dead season" of six months or more when rain-fed cropping is impossible. Rain-fed agricultural production is prey to the vicissitudes of the weather, and on-farm investments often reflect drought risk.

In addition to agriculture, the weather has a bearing on many other aspects of human life in Niger. Birth rates fluctuate year-round, achieving seasonal peaks during the months of August and September (Dankoussou et al. 1975). Humans have further adapted to seasonality and drought risk through spatial variation in land investments and income diversification (Watts 1983a, 1987; Mortimore 1989). Forms of adaptation are documented in ethnographies and oral histories (Smith 1965; Hill 1972; Cross and Barker 1991; S. Thomas 1992), which emphasize the use among the Hausa of space-spanning networks as well as of hierarchical social organization to adapt to risk and uncertainty. These habits are shared by other cultures within the ecological region of the Sahel.

The Sahel is therefore an ideal ground for such circular mobility systems to develop. Sally Findlay's research in Mali found migration of various durations, including a short-cycle, annual, off-season type lasting seven to eight months and a long-cycle variation lasting from two to seven years (Findlay 1992). These migrations are an immediate and long-term response to the threat of recurrent droughts. Short-cycle migration seems to have a close connection to periodic drought, especially among the poorest of the individuals whom Findlay sampled. Data collected in 1982 and again in 1989 found that, contrary to expectations, permanent migrations during a drought period did not rise. What changed was the proportion of total migrations that were short-term. Findlay also found corresponding increases in child fostering during droughts. The prevalence of short-term migration among women and children seemed to be an indicator of a drought's severity, since male circulation was high anyway.

Michael Mortimore interviewed dry-season migrants in Kano, Nigeria, in the midst of a drought in the early 1970s (Mortimore 1989). Eighty-four percent of the migrants had left land in the care of their parents or other relatives, 13 percent identified themselves as landless, and 3 percent had sold their land in the last two years. A majority of the migrants were married with children but had left their families at home, but 16 percent brought their families along with them. Ninety-three percent had access to income in Kano, but more than a third admitted that this income had come primarily through begging. Nearly all the migrants planned to return home. From these interviews Mortimore concluded that despite several years' harvest failure, there had been very little permanent redistribution of the population of northern Nigeria. Mortimore identified four strengths of the seasonal circular mobility compared with permanent migration: First, seasonal mobility was more flexible in terms of time and money; second, it was more adaptable to changing requirements of the migrant's life cycle; third, it was more flexible in terms of destination; and fourth, it could be conducted with little risk to property, since access to land and labor resources could be retained while absent (Mortimore 1989).

Flexibility, adaptability, and risk-reduction are all key attributes of seasonal, circular mobility, especially in drought-prone places like Mali and Niger. According to Ellen Taylor-Powell, seasonal migration has always been a fundamental feature of Sahelian economy and society and is "perhaps the principal income diversification strategy, and the most lucrative" (Taylor-Powell 1992, p. 17). The extent people go to to surpass the constraints of distance is a notable regional feature.

Given the setting for population mobility and the varieties in its uses, there are many complex societal interpretations that filter through individual and group behavior and bear heavily on observed practices. For instance, in anthropological fieldwork on Hausa mobility, Harold Olofson (1985) identified twenty-five emic (that is, locally defined) categories of spatial movement, all but one of which were circular. In a cosmological view of mobility, destinations are frequently indeterminate and little qualitative distinction exists between *yawon ganin gari* (walk of seeing the town) and *yawon ganin duniya* (walk of seeing the world).[8] In other words, there is not much cognitive or linguistic difference between stepping a few yards from one's door and traveling hundreds of miles away.

One can conclude that the differences between short- and long-term migration are governed by cultural norms and economic logic, but particular decisions to move are difficult to quantify owing to the flexibility of the practice.

To incorporate more cultural dynamics in mobility theory, the notion of migration itself needs to be reevaluated. Current conceptions are too strongly wedded to positivist behavioral traditions, with the stimulus of higher income yielding the response of migration. Emphasizing the stresses—the "pushes" and "pulls" of origin and destination—neglects the way the individual formulates and deals with these stresses. Instead, migration analysts should look at alternative conceptualizations that emphasize how potential migrants are situated within everyday life, which would involve a shift from a cognitive paradigm to a contextual paradigm. Gino Germani (1965) advocates the establishment of causation on three different levels: the "objective" level, using structural and community-level variables; the "normative" level, reflecting group attitudes toward migration; and the "psychosocial" level, related to the individual's motivation to move. To Gerald Haberkorn, "mobility is neither a unique result nor a unique cause of social, political and economic change but a reflection of an ongoing interplay between the general structural setting, people's more immediate physical-social environment of family and household, and their perceptions, interpretations, and actions" (Haberkorn 1992, pp. 816–817).

Thus we see multiple theories, none of which addresses the impetus of the individual. The *traditionalist* view that links contemporary migration to early forms of mobility, the *transitionalist* perspective that regards population movements as a temporary reflection of a country's continuing process of economic development, and the *dependency* explanation that contemporary mobility is the result of an uneven impact of capitalist penetration all fail because they downplay the potential of individuals affected by such changes to interpret and respond to changing realities in many different ways (Haberkorn 1992). Linking individual migrants with the environments in which they live and move requires juggling multiple contextual explanations, occurring on different spatial scales and at different temporal rates. Achieving such a linkage necessarily involves theorization.

Yet seasonal migration research continues with few stronger theoretical bases than these noted. Migration decisionmaking appears to be individual, and therefore *can* follow an income-based neoclassical framework. Clearly the context of the individual plays a role as well. How can we approach with any fidelity the migrant's actual decisionmaking process? Theories fail in this regard because they either essentialize or randomize human behavior, giving us, in Ronald Skeldon's words, only "the geometry of the process" (Skeldon 1990, p. 121). Linking so-called push factors to existing local environmental conditions requires a better understanding of place-based history and geography. Ways of adapting to change are built in part upon culture.

What role do the migrants' worlds—which include their cultures, social networks, and environments—play in mobility decisionmaking? Given the abundance of typologies, theoretical approaches, and perspectives on mobility activity in Niger and the rest of sub-Saharan Africa, it is helpful before proceeding to clarify our understanding of some basic propositions. Along with births and deaths, human movements are considered a primary component of demographic analyses. As a study of a dimension of human behavior, demography does not dwell in a value-free universe but depends on human ideas and actions occurring in specific settings. As a component of demography, mobility is subject to the same requirements. However one defines the environment, human actions are influenced by the setting or the conditions of the behavior. Therefore, understanding mobility itself cannot proceed without understanding its environment.

In weaving culture and environment into the fabric of mobility decisionmaking, we must also be prepared to accept another possibility: Perhaps people migrate for no reason other than that they simply like to. In the Sahelian region, attempting to link hypotheses of environmental degradation and population growth to the prevalence of time-worn activities is bound to be fruitless—even foolish—at times. In the interviews I did in the villages and towns of Niger, to the question why the men go away if there is no drought or apparent land shortage, I often received the enlightening reply that some people "just like it better going away." Their friends do it and it has become a tradition. A man whom the villagers in Guidan Wari knew went away and made enough money to start a small store in the village after he returned. The brother of an acquaintance from a nearby village went away and earned enough money to buy *two* oxcarts. As I have pointed out, eating and migration enjoy a first-order linguistic relationship in the Hausa culture of mobility. The heroism of circular migrants, and the fact that their activity is practically an initiation rite, mean that the exploits of migrants who *ci duniya* (eat the world) are legendary among young men. (See Photo 1.4.)

An understanding of the culture of mobility cannot be built on social, demographic, or remotely sensed data alone. Demographic analysis can help differentiate situations based on long-standing cultural norms from those that have been recently altered, but it cannot explain the multidimensional nature of regional change. Instead, we need a more panoramic perspective that includes the social dimensions of the rural-to-urban transformation, and this requires delving into complicated issues of cul-

Photo 1.4   Living the life: A man and boy pause from repairing flat tires at the main taxi-park of Maradi to pose for a photo. Few jobs in the city are not in some way touched by the culture of the road.

ture. Elements of a general framework have been put in place to explore the contexts of circular mobility by incorporating historical land uses, population and settlement histories, and degrees of interaction between settlements within the city and its periphery.

In Niger, population mobility flavors the entire economic and social life of the country. Explaining how circular labor migration in Maradi came about requires a more inclusive conception of mobility that challenges the dichotomies of traditional versus modern, internal versus international, rural versus urban, and push versus pull. Simply stated, to understand the connections between regional change and migration change in Africa requires us to put movements in their place.

The point is that circular mobility *persists,* as a culturally ingrained and culturally sanctioned response to hardship in a politically and environmentally capricious region. In the transformations occurring in West Africa, geography matters. Clearly the various types of mobility practiced here have developed under certain conditions and in certain environments over long periods, with changes in prevalence governed not only by cost and benefit calculations but also by the social uses of the mobility practices and by cultural and normative attitudes. The interactions between mobility and regional change are multifaceted and interlinked, and they are best approached on regional and local scales of analysis. To delve more deeply into the larger implications of individual movements, we need to understand the contexts of these movements as they develop on parallel and intersecting paths. To achieve a grounded population geography, we would do well to superimpose the mover and the mover's world, so that both the individual particularities and the universal connections—the environment of the movements, the world of opportunities that the migrant accesses, and the ways that the picture may be changing—become apparent.

Notes

1. "Eaters of the dry season" is a translation of the Hausa term *masu cin rani,* referring to those who look for food or those who are able to eat during the dry season. Polly Hill characterizes *cin rani* as dry-season migration undertaken to "eke out grain supplies" (Hill 1972, p. 217; see also Prothero 1957). The term *masu cin rani* is derived from the plural form of *mai,* "the owner of"; the verb *ci,* which means "to eat," "to consume something in its entirety," or "to win, overcome, or make the most of something" (Newman and Newman 1977); and the term *rani,* which is the Hausa word for the dry (or dead) season.

2. In 1992 the UNHCR estimated the number of registered refugees as 18.5 million and the number of displaced persons (those who are uprooted by force but remain within their own countries) to have been 18 million in Africa alone.

3. Questions have been raised about what makes "environmental refugees" "environmental." According to Astri Suhrke, for a migration to be environmental, it must be rooted in alterations in production or exchanges with nature that have particular specific ecological consequences (Suhrke 1991). In order to have any meaning, the notion of environmentally induced migration must refer to something more specific than those environmental changes that tend to be loosely associated with the uprooting of people. Otherwise it could not be separated from, for example, warfare-related refugee-creating situations. In fact, migration is rarely *not* environmental when it is related to the productive potential of a specific economic system or place and to the accumulated assets an individual has in that place. According to Jo Ann McGregor (1994), migration is usually one of a variety of survival strategies pursued by families either simultaneously or in sequence with other strategies such as selling assets, working for wages, eating bush foods, or undertaking short-distance migration.

4. It is unclear whether all the variants of mobility will increase at similar rates. Often the distinctions between "temporary" and "permanent" migration are less clear in actual fact than the terms suggest, with "permanent" migrants simply being "temporary" migrants who decide to stay longer in their destinations but never make a clean break with their place of origin.

5. One notable exception to this was the clearheaded analysis by Australian demographer John Caldwell (1975).

6. A note about Hausa orthography: Owing to the dueling British and French colonial traditions in Hausaland, Hausa words are conventionally spelled differently depending on which side of the border they originated, and some inconsistencies inevitably result. For instance, the latter-day descendants of the people of Katsina are called "Katsinawa" in the British spelling and "Katsinaoua" in the French spelling. I have tried to remain faithful to each term's origin: If a term is from what is known as Nigeria today, then I use the British orthography, and likewise use the French for a term originating in what is Niger today.

7. Pseudonyms have been used to protect individual migrants' identities.

8. It is noteworthy that *yawo* is often translated as "wander" as opposed to "walk," implying a less definite destination.

2

The Walked-Across Land

Ramatou is happy to talk to us while she dyes reeds for weaving mats. An elderly woman, she remembers her girlhood in the 1930s when the French occupied her village of Guidan Wari. French officers would pass through the village on their way from Tahoua to Tessaoua, riding large horses as "white as lamb's wool." The officers would walk up and down the streets looking at people but not speaking. They spoke no Hausa. Sometimes they would enter a woman's house and stare at her. Ramatou recalls when the French wrote down the names of everyone in the village and touched all the children.

This memory of a colonial census in the 1930s illuminates for us the way that historical events often appear to ordinary people. Africa's marginal position in the world, and the fact that it arrived on the Western scene late and could not set the terms of its participation in the Western world, makes history here unavoidably ironic, because here is where many old ways continue. History is too often presumed to be frozen in time prior to being recorded, a benchmark that usually corresponds to the start of the colonial era. But this point-of-encounter fallacy should not hamper our understanding why people originally came to settle along this environmental margin and why they continue to be mobile. Settlement history must go farther back and use every available method to derive a truer picture of how population and environment have interacted in place through time.

How do "conditions" affect the tendency to migrate? The phenomenon of seasonal mobility in West Africa is a feature of the Sahelian environment and climate: It is bone dry nine months out of the year here. A "walked-across land" in both historical and contemporary accounts, Niger has been the setting throughout the centuries for large-scale population movements, which have rarely resulted in dense settlement patterns (Dankoussou et al. 1975). Without revisiting the process of settlement, both before and during its inception, there is no way to know how the inhabitants' circumstances affected their mobility.

Being social as well as spatial, the condition of marginality occurs through the historical actions of specific individuals and communities, in phases that span the precolonial, colonial, and postcolonial periods. Case studies depict the poor as being pushed off their land by economic development or commercial activities that deprive them of their entitlement to resources necessary for their livelihoods. But other peripheries also exist, settled by those following the lure of uninhabited territory without the means to assess the long-term consequences of their land uses or by those who had fled religious or ethnic persecution and located themselves simply where land was available. Later-settled areas are likely to be subject to migration because these areas are environmentally marginal. The idea of migration is firmly established in the minds of people in these areas, and people here are likely to have active contacts elsewhere. In these circumstances, incorporating settlement histories and locating sites of dislocation allow us to relate contemporary movements to the ways the sites had been originally settled and effectively to play the animated film of human movements back to the conditions prior to settlement.

Settlement Dynamics and Population Mobility

African society, economics, and environment create patterns of settlement and agricultural land use that are characteristic of the continent south of the Sahara, with a low density of population and dispersed settlement patterns and with large areas of potential agricultural and pastoral lands that are unavailable because of eradicable diseases. These areas could be made more productive by a larger population engaging in more sustainable land uses (Turner, Hyden, and Kates 1993). The relationship between population and the environment along the margin of the desert is marked by a paradox in many places: The population density is too low to allow elaborate markets to develop but too high to enable the age-old systems of production to continue unchanged.

The distribution of the human population in Niger stems from both physical and human factors (Dankoussou et al. 1975). Physical factors include the location of water and the probability of precipitation, with population density correlating directly with amount and regularity of rainfall. Even with the phenomenal growth of cities in the latter half of this century, it should be remembered that the Sahel is still mostly rural. The population of the city of Maradi still represents less than 10 percent of the total departmental population (Ministère du Plan 1993). Settlement patterns in Niger have been determined by the presence of water. Most peo-

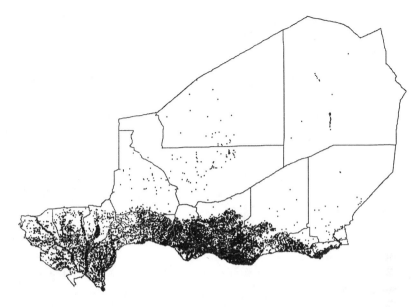

FIGURE 2.1   Population Settlement Dot Map for Niger

ple live in the southern agro-ecological zones, where levels of precipitation are high enough to permit crop agriculture during most years. The map of settlements based on the 1988 national census (see Figure 2.1) illustrates the macrolevel effect of precipitation and water availability on village locations.

Under the conditions of seasonal and highly variable precipitation, human settlement does not occur in a uniform fashion like a wave sweeping across a beach. The evenly spaced settlement pattern of south-central Niger is determined not only by the availability of water but also by social and economic geography and by history. Some locations in the south-central region, particularly in the Sahelo-Sudanian south, have been occupied for many centuries. Archaeological research has shown that the region west of Maradi near Madaoua has been populated continuously over the past 2,000 years, with cultivators using similar cropping patterns and techniques as today (Faulkingham and Thorbahn 1975). It is very likely that births and deaths in the population were balanced before the modern period, despite oscillations caused by temporary increases and then declines owing to famine. Other areas within the region were settled much more recently. In Niger, the prevailing farming practices—using short-handled hoes to cultivate millet and sorghum in outlying fields and growing cassava, cotton, and other higher-value crops closer to the village (Grégoire and Raynaut 1980; Brasset, Koechlin, and Raynaut 1984)—require abundant land and widely spaced villages, with uninhabited *daji* (bush) in between for grazing livestock and hunting and as a repository for medicinal plants and famine foods.

As in many places in Africa, the conditions for settlement in Maradi formed initially out of conflicts (David 1969). Episodes of warfare and consolidation efforts by rulers uprooted and displaced populations. The dislocations of these episodes were compounded by climate oscillations. During interwar or interdrought times, stable and sustainable long-term land uses could be established, but then they would be disrupted again by episodes of conflict and climatic variability. In Maradi, large swaths of land, with uninhabited lands in between, were settled rapidly as refuges between episodes of conflict. Only later did the spaces in between fill. In the northern part of Maradi Department, where the Sahelian agro-ecological zone lies, the isohyet[1] of 300 millimeters marks the boundary where the rainfall is thought to be too variable to allow permanent agriculture. North of this limit, which corresponds roughly to the fourteenth parallel,[2] the country was empty of settlements before the colonial period (Fuglestad 1983).

Guidan Wari is located in the western-central part of Mayahi *arrondissement* in the northern reaches of Maradi Department (see Figure 2.2), a "periphery within a periphery" of southern Niger. In its climatic characteristics, Mayahi is typical of the Sahelian zone, receiving perhaps 30 to 40 percent less rainfall than the city of Maradi, with a substantially greater drought risk (see Photo 2.1). According to the Health Ministry, Mayahi is the least covered by modern medical facilities of all of Niger's thirty-six *arrondissements*, with less than 12 percent of the population within two hours' walk of a clinic or hospital.[3] Mayahi *arrondissement* also has the largest percentage within Maradi Department of villages with chronic food deficits (Direction de l'Agriculture 1996). And Mayahi was the second most common source, after Dakoro *arrondissement*, of circular migrants in Maradi Department.

The village of Guidan Wari lies on the western edge of Mayahi *arrondissement*. Named in Hausa after Wari, its first chief, Guidan Wari (House of Wari) was founded in approximately 1890, a few years before the French began colonial rule, when sedentarized Bougajé pastoralists and Hausa farmers began to settle in the area (Jones 1995). With the exception of one period, the village has been ruled by the same family since its formation. Its 1995 estimated population of 968 is ethnically mixed, consisting of about 40 percent Hausa and about 60 percent Bougajé.[4] Guidan Wari has always been both Hausa and Bougajé, although many of the surrounding villages and the *quartiers* of Guidan Wari are predominantly Bougajé. Hausa is spoken by nearly everyone in the community, but there are families that speak both Hausa and

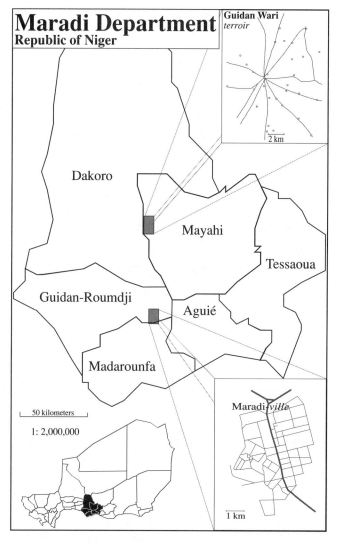

FIGURE 2.2 *Maradi Department*

Tamachek (the Tuareg and Bougajé tongue) in their homes. Guidan Wari is a Muslim village, but it retains some animist practices, including *Bori* (spirit possession).

The distance from Guidan Wari to the *arrondissement* capital, Mayahi-*ville* (the densely settled part that forms Mayahi proper), is 85 kilometers, which takes about four hours by bush taxi. The distance by paved and la-terite road to the department capital, Maradi, is 160 kilometers by way of

Photo 2.1   Guidan Wari occupies an ecological region that is bone dry for about nine months out of the year and has only been settled completely in recent decades.

Mayahi-*ville*, or 95 kilometers by sand path, either of which takes about eight hours by bush taxi. In 1992, a new laterite road was constructed connecting Guidan Wari with the cities of Mayahi and Maradi. The potential for a local market is still largely unrealized, especially for fertilizer and oxcarts, which villagers see as indispensable now that the bush and fallow lands surrounding the village have been all but totally consumed by agricultural fields.

Circular migration to and from Guidan Wari can be broken down into three basic types. Market migrants travel from Guidan Wari to Maradi and normally stay one or two nights in the city, in order to buy or sell at the Maradi market (Guidan Wari's market day is Friday). An estimated fifty to eighty of these men and women migrate from Guidan Wari each week, and this means representation by approximately 10 percent of the households of Guidan Wari. Seasonal migrants sometimes consist of families who leave the village after the harvest and head toward urban centers in Niger or in other countries within Africa to look for work. In Guidan Wari, about ten households migrate in this way, usually taking up residence in Maradi or in Jibiya, across the border in Nigeria. The duration of such movements is usually around six months, after which time the migrants return to their villages. Temporary migrants are usually men who seek or resume livelihoods on a short-term basis in the city, and about thirty migrate per week from Guidan Wari itself to Maradi and points south for this purpose. Most of the temporary migrants do not actually come from Guidan Wari itself; rather, they come from smaller surrounding villages. In these satellite villages, which were settled later than Guidan Wari, residents estimated the various kinds of circular migration to affect about 50 to 75 percent of households directly. These estimates exclude those who have left permanently.

I conducted group interviews with villagers in Guidan Wari and eight surrounding villages on changes in the local environment, including the disappearance of flora and fauna, changes in the areal extent of farming and farming methods, drought histories, population change, the connection between village growth and the loss of bush land, and their effects on the prevalence of circular migration. While in Guidan Wari, I heard and recorded a variety of perspectives on circular migration from men and women in the village. My time in Guidan Wari was particularly fruitful in illuminating just how divided the community has become over subjects related to mobility. But I stumbled time and time again over the subject of population.

How did Guidan Wari get to be like this? Where did everybody come from? Measuring numbers of people alone has less explanatory power than the geographical particulars of where people live and how they move. Associating the birth rate with the prevalence of circular migration is meaningless unless cast in terms of a community's ability to satisfy food needs, which I argue cannot be done without looking at the nature of settlements, specifically how and why they formed. For the Sahelian agro-ecological zone in the north of Maradi Department, the role of climate, conflict, and colonialism are the underlying factors that, in concert with changes in land use and political economy, are responsible for today's rates of seasonal migration. These are far more meaningful as explanations than the simple arithmetic of population growth alone.

In the quest to incorporate conditions in demographic analysis—conditions that have a bearing on the relationship between settlement dynamics and mobility—we must ask when and under what circumstances communities formed. The *order* of settlement affects the access to resources such as farmland, water, and markets, and these access dynamics can only be seen on the spatial scale of the *terroir* (village group). For assessing present-day vulnerability in Maradi Department, knowing how the region was settled matters more than knowing the size of the contemporary population using its resources. Marginal or latter-settled areas are often incapable of maintaining food sufficiency regardless of the size of the population. The "problem" is the temporariness of the movements themselves and their documentation.

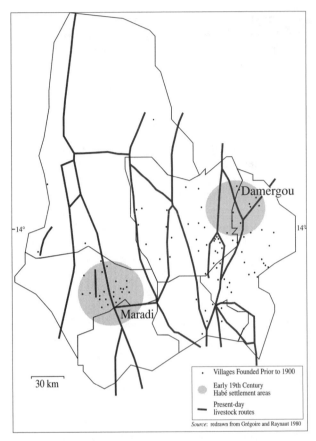

FIGURE 2.3   *Livestock Routes and Pre–1900 Settlements, Maradi Department*

Precolonial Settlement History of
Maradi Department

Prior to the era of conquest, which lasted until the end of the nineteenth century, the land uses of the Sahelian zone were mainly pastoral, as is still evidenced by modern maps of livestock corridors (see Figure 2.3). The early economy was based on age-old interactions between grain-growing sedentary farmers to the south and livestock-herding nomadic pastoralists and on long-distance trade between the desert edge and the cities to the south (Baier 1980). Precolonial movements were mainly north-south, that is, perpendicular to the rainfall gradient; later colonial and postcolonial movements were oriented more from east to west.

The cultural history of the inhabitants of the region reveals divergent processes, including the sedentarization of the pastoral population and

Photo 2.2 Settlement of this region is marked by northward movement of farming colonies and sedentarization of the herding population. This scene at a busy cattle well underscores the area's pastoral base.

in-migration from the south. Tuareg and Bougajé pastoralists in the Sahelian zone of Maradi Department are thought to have altered their age-old seasonal herd movements in the nineteenth century. When asked why the Bougajé stopped their nomadic herding and settled down, the elders in Guidan Wari replied that the Bougajé had never been completely nomadic anyway and that even their grandparents were more or less settled (see Photo 2.2). No one knows if people lived in the area around Guidan Wari before the village was settled. The Hausa agriculturalists who moved into the Sahelian zone of Maradi in the nineteenth century are known to have come from the south, and they used techniques that originated and were mastered in the Katsina and Gobir states of the Sokoto Caliphate in the Hausa cultural hearth. Located in the Sudanian agro-ecological zone, Katsina and Gobir receive substantially more rainfall than the Sahelian zone of Maradi.

The historical context for settlement of this Sahelian zone of Hausaland was researched by Jim Delehanty (1988), who documented the peopling of the farming frontier through family fragmentation and northward movements in the 1920s and 1930s. He identified a frontier ethic at work in the settlement of Gangara in Tanout *arrondissement* and also pointed out the roles of social and political conflict. The settlement patterns of the precolonial frontier fringe depended on the location of polities, or forms of local political administration. Farmers lived on the periphery in precolonial times because they decided that obligations that went with fealty to a distant power were not worth the protection the power afforded (Delehanty 1988). Along with the expansion of Tuareg control over the transition zone of central Niger after the fifteenth century had come the removal of farming, as the Tuareg *kels* (clans) drove farmers out of the Aïr Mountains and created a no-man's land in the Sahelian zone. This land remained empty for centuries (though it continued to be used for grazing livestock) through the policies of precolonial political entities. Borno-controlled Damagaram (present-day Zinder) in the nineteenth century arranged a farming limit with the Tuareg (Salifou 1971), which effectively kept out farming settlements.

Conflict can preserve environmental variety, and environmental depth can mitigate the effects of conflict. By concentrating the flows of people in some areas and making other areas off-limits, conflict effectively preserves some areas' environmental inventories. Nature also provides hiding places for outcasts and refugees of conflict. In Maradi, for instance, religious warfare initially inspired the settlement of the region, and later the cessation of conflict opened up areas to conquest and colonization. Accounts from travelers confirm this relationship. On his trek through Hausaland roughly fifty years after the Muslim uprising of the Sokoto jihad, Heinrich Barth (1857) observed thick, intact forest in disputed territory and in vast areas that had been abandoned years before because of religious conflict.

Barth's account helps re-create the scene from the early nineteenth century, when the Fulani led by Usman dan Fodio were fighting a religious war, starting in 1802, to consolidate the seven Hausa states. One effect of the Sokoto jihad was to detach the sparsely settled northern fringes from the rest of Hausaland. Usman dan Fodio's Sokoto Caliphate created an indigenous frontier between *Habe* animists in the north and the Fulani zealots of the jihad in the south. Around 1819, the exiled *Habe* king of Katsina reestablished himself as the independent ruler of Maradi, in the northern provinces of his former realm. In the Maradi-Tessaoua region, the thick forest that had previously covered the Maradi valley was partially cleared for farmland (David 1969).

The settlement of the valley enabled the new state of Maradi to acquire its separate status, which it used as a base to attack Sokoto. Twenty years later (in 1839) the same sequence of events occurred at Tibiri, where the exiled king of Gobir established himself as ruler (Thom 1975). A Katsinawa chief established an additional base at Tessaoua. The Goulbin Maradi valley, as well as the Sahelian northern zone of Hausaland, including the Goulbin Kaba near the arc running from Mayahi to Tessaoua, and the hills of Korgom southeast of Tessaoua, became refuges for the exiled victims of the Fulani against the incursions of Usman dan Fodio's zealots, who were established in Katsina, Kano, and Sokoto. For almost a century, the struggle continued between the Fulani-dominated Katsina and Gobir and the descendants of the traditional *Habe* rulers, bringing devastation to the area. Barth's map of this early Maradi state features a forested region, the noted "northern limit of cotton" (*Nordgrenze Baumwolle*), and a territory inhabited by "unassimilated heathen tribes." The jihad triggered sizable migrations to the north as well as to the south. In this region of refuge in the northern reaches of Hausaland, cotton and millet were cultivated in detached, widely spaced villages for much of the nineteenth century. As it is today, this Sahelian zone was subject to fluctuations in rainfall. In a description of a village located at the fourteenth parallel, Barth describes a watering place "consisting of a group of not less than 20 wells, but all nearly dry. The district of Damerghú must sometimes suffer greatly from drought" (Barth 1857, p. 427).

This recounting of the precolonial history of Maradi Department should make clear that the northern reaches of Hausaland are a collection of borderlands that were settled as refuges from conflict, peripheral to the main centers of power in the nineteenth century. Climate has also played an important role in the settlement of Maradi's Sahelian zone. Village locations have been determined by the presence of water. Most people live in the southern agro-ecological zones, where levels of precipitation are high enough to permit crop agriculture during most years. To Finn Fuglestad (1983), the late settlement of the northern parts of Maradi suggests that the sedentary societies of the region are less articulated and more anarchistic than nearby societies and that many areas within the frontiers of present-day Niger were probably very thinly populated, and contained vast wastelands, until the nineteenth century. The third factor, and perhaps the most influential, in the settlement of Maradi's Sahelian zone is colonialism.

Settlement in the Sahelian Zone
During the Colonial Era

Maradi's potential must have attracted the European conquerors of the time, because British and French competed heartily for the territory of northern Hausaland in the 1890s, and the line separating the areas they claimed was redrawn no fewer than four times between 1890 and 1906 (Thom 1975). The line was originally drawn from Say to Baruwa in 1890, putting Maradi in French-held territory. After the Anglo-French Treaty of 1898, the line followed an arc about 160 kilometers from Sokoto until it hit the fourteenth parallel, leaving Maradi-*ville* within British territory. The Third Military Territory of Niger established its headquarters at Zinder. To travel west to the Zarma-Songhai region, there were two routes cavalry could use: a *route désertique* from Zinder to Filingué via Tessaoua, Tahoua, and Bouza, which was sandy, treacherous, and devoid of water holes, and a *route practicable,* which after 1898 ran through British territory and required British escorts. The tensions between Britain and France over these access routes led to a settlement in 1904 in which the repeal of French fishing rights in Newfoundland was traded for the *route practicable* across the French colonial territory (Thom 1975).

This delimitation was significant for the balkanized states of colonial Africa because, for the first time, European powers used precolonial demarcations as a basis for the creation of a new boundary. The precolonial states of Mantakari, Adar, Maradi, and Tessaoua all fell under the control of France, and to the British went Sokoto, Zamfara, Katsina, Daura, and Bornu. The remaining states of Maouri, Konni, Gobir, and Damagaram were divided by the boundary, but of these states, France received the bulk of the territory. The actual subdivision was undertaken in 1908 by Captain J. Tilho with an expedition of ethnographers, botanists, and cartographers. The British regarded the French curiosity in its new territories as a delaying tactic. The map that resulted from the expedition shows the towns of Tibiri, Maradi, and Tessaoua and a chain of villages along the Goulbin Kaba, which was then a seasonal river bordered by doum palms, mimosa, and recessional agriculture. Vast empty spaces and abandoned villages marked the transition zone.

The peripheral areas of northern Maradi did not react immediately to the arrival of the French. Having struggled with the British for control of the West African interior, the French government back home was not wholly willing to finance its development.[5] The commercial apparatus established in Maradi-*ville* in the 1920s, with trading houses and new forms of inventorying and extending credit, was intended to support a low-budget colonial administrative presence, and a scheme introducing the groundnut was designed to gain revenue.

Whatever its budgetary constraints, the French set out to dominate their *territoire militaire.* From 1906 to 1915, hut taxes tripled and had to be paid in French francs only, with cowrie shells no longer accepted as currency after 1916. Forced labor was requisitioned at French bidding (Fuglestad 1983). A consequence of these draconian policies of the early colonial administration was the southward migration of peasants out of French territory and into British-ruled states, which had more lenient policies. These movements were attributed to "the heavy burden of the French occupation" (Lugard 1902, p. 34, as reported in Thom 1975). The French reacted by attempting to stop the population movements, even of the nomadic Fulani, by delimiting grazing zones and corridors of transhumance, and erecting customs barriers along the frontier. After it became clear that the revenue collections were not having their desired effects, the policies of hut taxes and compulsory labor were suspended, but not before new settlements were established between Katsina and Daura and between Maradi and Zinder as a result of southward migrations. The population movements, as well as the redrawn colonial boundary, affected the location of markets, particularly the border market at Jibiya.

Another consequence of colonial policy was the outward expansion of populations from precolonial refuges in the Sahelian zone, an expansion brought on by the release of thousands of slaves after 1906 (Fuglestad 1983). At first the French were indifferent about the need to end the use of black slaves by the Tuareg, but a federal decree issued by authorities in Dakar in December 1905 stipulated heavy penalties for all forms of slave trade and signaled to officers in Niger the necessity of taking action. In June 1906, a *circulaire* (bulletin) was released announcing that slavery was incompatible with French rule. This decree, combined with the defeat of Kaossen, a legendary Tuareg leader, marked the end of Tuareg hegemony over central Niger and the beginning of the period of sustained northward migration by Hausa farmers from the south.

Villages were formed throughout Maradi Department in the first decades of the twentieth century, partially in response to French colonial policies. In 1917, there were few Hausa villages north of the established Hausa states. The Hausa farmers who had lived in the north prior to the French colonization had not paid tribute to the Hausa kingdoms, making them Hausa by custom but not by political allegiance (Delehanty 1988). In 1926, the French geographical service of the army classified vast parts of *le Cercle de Maradi* (the area under Maradi's administrative control) as *"régions inhabitées* [uninhabited regions]" or *"semi-désertiques,"* notably the area between Maradi and Aguié, the land reaching the Nigerian border and almost all the land under the line from Kornaka to Ourafane, which runs slightly north of fourteen degrees. Farmers moved progressively as far north as the Tarka Valley. Of the 600 villages studied by Projet Maradi in the late 1970s, 60 percent were founded between 1900 and 1929.

By the 1920s, expanding markets for groundnuts, cattle, and millet in northern Nigeria and the introduction of a new hoe led to increased yields and further land conversion. The apparent shortage of land in the south induced many farmers to move into regions hitherto considered unfit for permanent cultivation, in the region north of the 300mm isohyet. The shortage of land also adversely affected the nomadic Fulani, pushing them into areas formerly controlled by the Tuareg. The Bougajé, reputedly former slaves of the Tuareg, were slowly integrated into the Hausa world and developed their own agro-pastoral economy. According to the very sketchy estimates done by the French, Niger's population increased from about 1 million in 1922 to more than 1.5 million nine years later, despite two severe epidemics of meningitis and typhus that killed an estimated one-tenth of the population between 1923 and 1927 (Fuglestad 1983).[6]

This region became occupied at the time of overall above-mean rainfall in the 1920s and 1930s, despite drought years of 1913 to 1915, 1926, 1931 to 1932, and 1937 in some areas. According to Fuglestad (1983), the French displayed great ignorance of the dynamics of drought: The ill-timed enlargement of French demands tended to coincide with periods of unfavorable climate conditions.

Each time the French raised taxes, the Hausa response was to produce more and to sell their surpluses on the market, and groundnut cultivation spread across the region. Road links east to Zinder through Tessaoua, west to Madaoua, and south to Jibiya were constructed beginning in 1921 (Grégoire 1992). Groundnuts were transported by camel and donkey to the Kano railhead until 1953, when Opération Hirondelle (that is, Operation Swallow) created a road from Zinder to Parakou (in Dahomey, or Benin today) to bypass Nigeria because of the high rate of spoilage owing to delays at Nigerian railways.

The increased tax burden and the resultant expansion of the extrasubsistence sector of the economy led to the clearing of new land, an activity made possible by the French control of areas previously left empty for reasons of security. Since concentration made people increasingly vulnerable to tax and labor demands and since many people disliked living in large villages because farmland was less accessible, settlements broke up in many areas of the region (Fuglestad 1983). A disastrous Vichy interlude from 1940 to 1945 featured the French requisition of agricultural surpluses and a failed French attempt to introduce the plow to Hausa and Bougajé peasants. The era beginning in 1946 was marked by the rise of the Fourth Republic, with African representation in the French National Assembly, during which Niger enjoyed higher groundnut prices and a favorable climate. These factors further encouraged farmers to convert more land from grain (subsistence crops) to groundnuts.

The result was an unprecedented expansion of the cash-crop sector of the economy after 1945. Groundnut production in 1948 amounted to 45,500 tons; by 1955, it was 70,000 to 90,000 tons. The acreage devoted to groundnut production in Maradi doubled between 1951 and 1960 (Pehaut 1970), partially at the expense of the subsistence sector. This expansion was fueled by French intervention. Price supports were in effect from 1954 onwards, as were subsidies to cut transportation costs and regulations for marketing. Groundnuts were being moved by truck from the mid–1950s onwards. All these incentives encouraged the small producer to grow groundnuts. The French did not provide food to farmers, although cash gifts were provided to participants during the rainy season. The consumer goods available for sale in Maradi-*ville* and through a network of merchants in rural areas included foodstuffs (sugar, rice, salt), trade items (lamps, bowls, fabrics), and construction materials (Grégoire 1992).

The conversion to groundnuts raised alarm among the colonial officers about the effect on famine vulnerability because of the loss of land previously devoted to millet and sorghum. Warnings about the dangers represented by the northward extension of cultivated land and the expansion of the cash-crop sector appeared frequently in official reports during the 1950s (Brasset, Koechlin, and Raynaut 1984), as evidenced in the 1957 issue of an annual report of the subdivision of Tessaoua. But apart from a decree in 1954 establishing a northern limit of cultivation, which was impossible to enforce, administrators did not heed the warnings and instead continued to encourage farmers to grow groundnuts. The tumble in groundnut prices during and after the 1950s, combined with the increase in taxes, meant that Hausa agriculturalists had to sell ever more groundnuts to maintain their standard of living. Colonial administrators noted the implications for famine vulnerability but did little to act on them.

In his analysis of the modern history of Niger, Fuglestad raises questions about the potential in Nigérien Hausaland for what is called in the West "economic development." Whatever money the Hausa people earned from selling groundnuts was apparently not invested in productive undertakings but, instead, was contributed to what are known as "consumption activities," mainly gift giving. To the French, the Hausa people were regarded as "traditionalists," and in comparison with the more compliant Zarma in the west, their loyalty to France was suspect, partially because of their connection to British territory. In Fuglestad's view, "The Hausa opposed the French with a particular kind of passive resistance, and a successful one from the point of view of the 'traditionalists,' since Hausa society remained relatively impermeable to French influence" (1983, p. 77). The economic system continued to be based not on the principle of accumulation but on the principle of permanent redistribution of wealth.

It should be noted that from the perspective of famine risk, this economic philosophy makes sense. People are more useful in times of drought than are money or things, and building a network to mitigate the effects of famine is more rational locally than accumulation. The devastating droughts that occurred eight years after the end of colonial rule, whose effects can be seen as a consequence of the settlement of marginal land, demonstrate this. Still, the fact that it was the ethnic group considered least contaminated by modernity that monopolized the cash-crop sector is "a paradox which remains to be properly investigated" (Fuglestad 1983, p. 178). In the ensuing independence movement and the establishment of the national capital in Niamey on the western reaches of

Niger, the Hausa were penalized for their aloofness and came to
inated politically by the Zarma.

We have seen in this extract from Maradi history the role of conflict, climate, and colonialism in determining the land settlement pattern of
Maradi Department. Conflict kept the Sahelian zone of Maradi Department very sparsely settled until well into the present century. Higher
than average rainfall in the 1920s and 1930s encouraged the northward
migration of farmers. Colonial policies such as forced labor and heavy
taxation encouraged settlements across the frontier to the south and fueled the northward migration of farmers. The introduction of the
groundnut as a cash crop and the establishment of a road and market infrastructure to service groundnut production fueled a land grab in the
Sahelian agro-climatic zone, which had formerly been regarded as un-
suitable for permanent cultivation. As we will see when examining at a
finer scale the closing of the farming frontier, new villages emerged after
the period of initial settlement as infill between older villages. Once the
land in this economically marginal zone was occupied, the stage was set
for later and different mobilities.

Regarding the relationship between population growth and intensification, it should be noted that access to markets, networks, and resources
can work both ways. Paul Richards theorizes that in the forest zone of
West Africa, population loss resulting from male migration *led to* intensification because traditionally the men clear the forests for extensive systems and their absence increased the importance of female labor to increase output on land already cleared for cultivation (Richards 1985).
Sara Berry points out the positive feedback that can exist between population density and market access when access draws immigrants into an
area and high population density encourages development of a market
infrastructure (Berry 1989). Given the gender division of labor prevalent
throughout West Africa, there is no reason to suppose that these reciprocal effects are any different in Maradi Department.

In marginal areas, land-based investments are low because risk is high,
and, as I have argued, this investment-risk relationship stems from settlement dynamics. In the frontier phase, the abundant available bush land
provided the necessary margin. On the Sahelian settlement frontier from
1930 to 1975, the land-use intensity[7] grew from approximately 5–10 percent
in 1930 to about 18 percent in 1955 and to about 45 percent in 1975 (Brasset, Koechlin, and Raynaut 1984). By the early 1960s, most settlements in
Mayahi had become fixed, with few if any new villages appearing. According to the planning office of Mayahi *arrondissement*, there are 343 villages in Mayahi; the frontier is officially closed (interview with Oumarou
Souley, December 1995). Without new land to exploit for needed food production in the years since the frontier closed, villagers have found them-

a tightening vise, with few recourses to cope with starvation in their villages. As the available options have shrunk, seasonal circular migration has emerged to allow access to urban cash, which is intended to close the gap in food supply. This transformation in the uses of mobility, from income supplement to food security guarantor, will be different in different places, but in general the increase in reliance on remittance income occurred simultaneously with the filling of unoccupied bush land. This process can be observed by analyzing population census data available for the region at the juncture when the farming frontier closed.

We must not forget that this time period represents only a snapshot in the long history of changing settlement patterns in Mayahi *arrondissement* and that the periodicity of this change may conceal other settlement factors. For instance, the surges of dislocated peasants southward in the late 1960s resulted in the depopulation of hundreds of villages, which recovered afterwards at rapid rates. There is of course no way to document accurately the growth rates of villages in the decades prior to demographic record keeping. In 1977, faster rates were recorded in the larger villages than in the medium-sized villages; among the smallest villages, high rates of growth are recorded. It is likely that this growth did not occur through natural increase alone but must include movements from smaller to larger villages and towns. Some of this movement could be explained by the availability of more off-season and extra-agricultural opportunities in larger villages, and since few new villages are likely to have been settled during this period, people are therefore migrating toward established settlements. Analyzing the intercensal village growth plotted by geo-referenced village location sheds more light. Figure 2.4 maps the annual growth of villages between 1977 and 1988.

To approximate the conditions of village growth in the period between the two censuses, a map of Mayahi's agro-ecological zones has been added to the analysis of intercensal population change. The data for the agro-ecological zone map came from Projet Maradi, which collected data on soil, topography, vegetation, precipitation, and land-use types and intensities, and categorized them into six major groups and twenty-four subgroups (Brasset, Koechlin, and Raynaut 1984).[8] Figure 2.5 is a map of agro-ecological zones for Mayahi Department.

To give a rough sketch of the different agro-ecological zones in Mayahi *arrondissement*, zones A (including A_8 and A_{11}), B (including B_0 and B_2), and C (including C_0 and C_1) are in the more arid and less densely populated parts of the Mayahi and were mostly still frontier at the time of the 1975 aerial survey. These zones are dominated by pastoral land uses and low population densities (five to seventeen persons per square kilometer), with some northern areas settled by Hausa farmers and Bougajé pastoralists in more recent times.

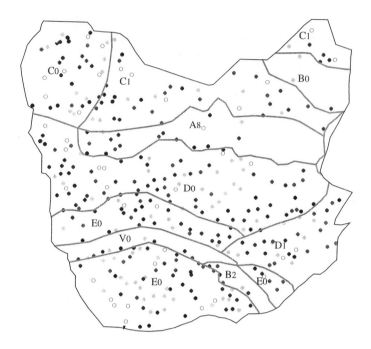

Annual village growth

- 7 to 32 percent
- 4 to 6.99 percent
- 3 to 3.99 percent
- ○ -0.9 to 2.99 percent
- ○ -17 to -1 percent

FIGURE 2.4   *Mayahi Village Growth, 1977–1988*

Zones D (including D_0 and D_1) and E (E_0) in the south were the areas first settled. They too are dominated by crop agriculture. These zones feature old refuges like Damergû in the D_1 zone, as well as some water holes and market towns along the old caravan routes. Subgroup V_0 is the Goulbin Kaba, the arc-shaped valley to the south of Mayahi-*ville*, featuring doum palms (Latin name *Hyphaene thebaica; kaba* in Hausa) and pastoral land uses. The doum palms in this valley might originally have been preserved to shield villages against slave raiding from the south by preventing horseback riders from entering and leaving the area rapidly (Grégoire and Raynaut 1980). Today the trees rarely grow more than a meter high before their palms are harvested. Traditionally settlements and agricultural fields have been kept out of the Goulbin Kaba to ensure its continued function as a livestock corridor, but recently some encroachment has

Agro-Ecological Zonation of Mayahi *Arrondissement*

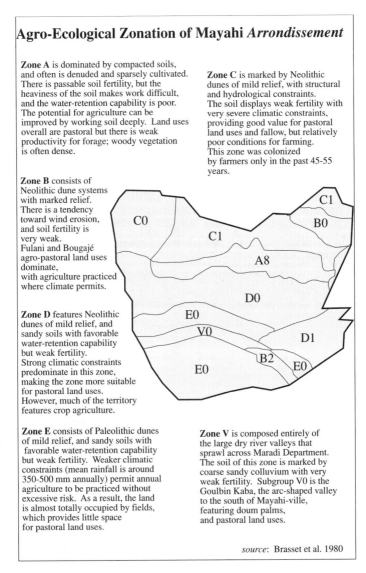

Zone A is dominated by compacted soils, and often is denuded and sparsely cultivated. There is passable soil fertility, but the heaviness of the soil makes work difficult, and the water-retention capability is poor. The potential for agriculture can be improved by working soil deeply. Land uses overall are pastoral but there is weak productivity for forage; woody vegetation is often dense.

Zone B consists of Neolithic dune systems with marked relief. There is a tendency toward wind erosion, and soil fertility is very weak. Fulani and Bougajé agro-pastoral land uses dominate, with agriculture practiced where climate permits.

Zone C is marked by Neolithic dunes of mild relief, with structural and hydrological constraints. The soil displays weak fertility with very severe climatic constraints, providing good value for pastoral land uses and fallow, but relatively poor conditions for farming. This zone was colonized by farmers only in the past 45-55 years.

Zone D features Neolithic dunes of mild relief, and sandy soils with favorable water-retention capability but weak fertility. Strong climatic constraints predominate in this zone, making the zone more suitable for pastoral land uses. However, much of the territory features crop agriculture.

Zone E consists of Paleolithic dunes of mild relief, and sandy soils with favorable water-retention capability but weak fertility. Weaker climatic constraints (mean rainfall is around 350-500 mm annually) permit annual agriculture to be practiced without excessive risk. As a result, the land is almost totally occupied by fields, which provides little space for pastoral land uses.

Zone V is composed entirely of the large dry river valleys that sprawl across Maradi Department. The soil of this zone is marked by coarse sandy colluvium with very weak fertility. Subgroup V0 is the Goulbin Kaba, the arc-shaped valley to the south of Mayahi-ville, featuring doum palms, and pastoral land uses.

source: Brasset et al. 1980

FIGURE 2.5   Agro-Ecological Zones for Mayahi Arrondissement

been occurring. Brasset, Koechlin, and Raynaut (1984) speculate that the E_0 zone, south of the V_0 zone, like many areas was settled in the 1920s and 1930s, but the E_0 zone filled early and appears on the 1975 land-use intensity map to have very little bush land available for conversion. The land in these earlier and more densely settled zones had already closed their agricultural frontiers by the time the French carried out their 1975 aerial photographic missions. These southern zones feature population densities ranging from thirty-one to forty-five persons per square kilometer.

The analysis of population change by agro-ecological zone, presented in Table 2.1, shows that between 1977 and 1988, population did apparently grow in zones with higher population densities, but with a great deal of variation in the rates. Zones C_1, B_0, and C_0 experienced negative population growth during the time period, whereas zones B_2 and E_0 and observed annual growth rates in excess of 5 percent.[9] Judging from this variation, migration has very probably occurred from areas with low densities to areas with higher densities, in other words, in areas that are already "filled." This sheds light on the role played by environmental conditions in directing the flow of migrants. The analysis of population growth in the E_0 agro-ecological zone from 1977 to 1988 reveals that the population has grown especially fast in an area that apparently has no more land,[10] suggesting that population growth in less-dense areas might continue in the absence of bush or fallow land conversion.

In the more densely populated southern part of Mayahi *arrondissement*, large villages are growing especially rapidly, and as the land-use intensity analysis will show in more detail in the next chapter, this growth has occurred even when there is little or no available land. In sparsely populated land that is less suitable for agriculture, the populations are growing the least quickly, with negative population growth experienced in three zones. Spatial variation in bush and fallow land—especially differences regarding change between north and south in land converted to agriculture—reveals the role of land shortage in creating the stress of a closing frontier. To fill out this context, we must delve to a microscale and examine the changes on the village level. To see the consequences of village growth with respect to the order in which the area was settled, one must scale down further to the village group, or *terroir*. If the farming frontier has indeed closed, then the spaces between villages have filled in. This lack of free land will prompt a transition in mobility types, from frontierward movements to seasonal and subseasonal movements toward urban centers.

Settlement and Population
Dynamics on the Local Scale

Investigating the dynamics of settlement on the level of the *terroir* illustrates how the dynamics of settlement order influence both the landscape and, by extension, circular mobility. The settlement stories and dates are based on field investigations done with the help of Emlyn Jones, a second-year Peace Corps volunteer who acted as intermediary and inter-

TABLE 2.1 Population Change by Agro-ecological Zone

Agro-ecological Zone	Area (km²)	Number of Villages	1977 Total	1988 Total	Average Average Size	Population Density (1988)	Population Change, 1977–1988 (annual)
B_2	103.62	6	822	1779	297	17.17	7.02%
E_0	1557.42	113	25690	51563	456	33.11	6.33%
V_0	253.37	9	1936	3163	351	12.48	4.46%
D_1	498.33	38	14145	22767	599	45.69	4.33%
D_0	1478.09	99	35835	46413	469	31.40	2.35%
A_8	663.33	17	6562	7673	451	11.57	1.42%
A_{11}	29.16	1	158	181	181	6.21	1.24%
C_1	1059.58	43	15544	15536	361	14.66	-0.005%
B_0	261.68	4	1358	1354	339	5.17	-0.03%
C_0	619.49	44	13184	10762	245	17.37	-1.85%
Total/average	6524.07	374	115244	161191	431	24.71	3.05%

preter, to demonstrate the processes and consequences of frontier settlement.

Guidan Wari consists of a group of family residences centered around two wells and a market, surrounded by fields. A typical family compound in Guidan Wari includes a number of straw huts enclosed by a millet-stalk or clay *banco* (mud-brick) wall, with each adult living in his or her own hut. The residential pattern is virilocal, with husbands and their wives living with the husbands' families in the village. There is a primary school in the village. About ten men, including government functionaries, speak French and are literate in French; about fifteen men are literate in Arabic; perhaps 10 percent of men and less than 5 percent of women are literate in Hausa (Jones 1995).

The traditional power structure of Guidan Wari exemplifies the advantages for the village of a strong *mai gari* or *chef du village* (chief), a position that includes responsibilities of dispute resolution, mobilization of labor, and tax collection. The degree of social cohesion or the power of the chief is a strong factor behind the success of villages like Guidan Wari. In Guidan Wari, the chief handles most of the affairs of the village, including annual tax collection from the villagers. Of all the chief's tasks, perhaps the most important is land allocation. The traditional and governmental hierarchy have a close relationship in this village.[11]

The *terroir* of Guidan Wari as a whole, within a circle of five-kilometer radius, has a population roughly estimated at 3,500,[12] making the estimated population density approximately 44.5 persons per square kilometer. This density is substantially higher than the average of 31.4 persons per square kilometer measured for the D_0 agro-ecological zone. Nearby villages included within this five kilometer circle include Guidan Wani, Dadin Tamro, El Sangia, Kollo, Guidan Boujia, Allah Karebo, Mallameye, and Guidan Antou. In nine villages (including the largest, Guidan Wari), I asked questions about settlement histories, population, land uses, and estimated prevalence of circular mobility. Driving forces of settlement that have been discussed in the context of the Sahelian zone have their counterparts on the village level, and this has affected the order of settlement in and around Guidan Wari from before the period of nineteenth-century colonization. Figure 2.6 shows the pattern of village settlement for Guidan Wari *terroir*.

In the precolonial period, settlement occurred through sedentarization of the pastoral population. In the late 1800s, the Bougajé (for reasons apparently even they themselves do not know and years before the French banned the practice of slavery) detached from their nomadic Tuareg overlords and settled down in a *zango* (enclave) in the vicinity of Guidan Wari. Villagers were also aware of the historical northward movement of Hausa farmers into the area. They were apparently drawn by the market

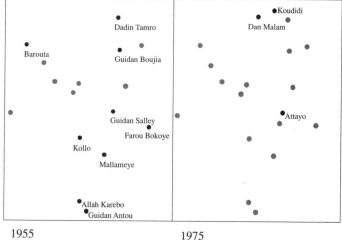

FIGURE 2.6   Dates of Village Formation, Guidan Wari Terroir

there, which was located on a critical livestock route. Hausa migrants came from the south by way of Alhassane, on the banks of the seasonal Goulbin Kaba, which flowed reliably 100 years ago. The earliest village founded by the Hausa settlers was Guidan Wani, located about five kilometers east of Guidan Wari.

With a geometric street layout and a compact Hausa feel to it, Guidan Wani was founded around 1889, although some maintain that the village is actually 150 years old. When the Guidan Wari chief's family settled, according to the villagers interviewed, Guidan Wani was already there. It is one of the oldest villages in the area. A cement well was built in 1968, which complemented a traditional well that is slightly older and still productive. Under the administrative control of Guidan Wani are Guidan Salley and Taraci, both of which were founded around 1940 by people originating in Guidan Wani. All together the villages have about thirty-five households, with a total population of perhaps 800. According to the chief of Guidan Wani, there are no newcomers in the village, and no new houses. The population here has grown by natural increase alone.

Within the *terroir*, Guidan Wari and Guidan Wani are the oldest villages, and in relation to newer villages they enjoy a privileged position on the land. Guidan Wari and Guidan Wani occupied the area for decades prior to the settlement of other villages and enjoyed the hunt and the ample wood supply while it lasted. Today they still have the better resources, including more wells, closer fields, and more oxcarts, giving them higher prestige within the *terroir*. They also had the advantage of already being there to witness the beginning of French colonial rule.

According to the elderly informants in Guidan Wari, the colonial era in the *terroir* began quietly and was marked by a steady stream of in-migrations and new village formation. The French indeed were an early presence here because of the *route désertique* connecting Zinder and Tahoua. The French had a garrison staffed by three or four officers, along with some Hausa-speaking *soji* (soldiers), living in the village in a house near the road. The census that the French conducted in the 1930s, which Ramatou remembered, was lost in the jumble of administrative archives in the city of Maradi.

Guidan Wari's founding date of about 1890 corresponds to the earliest contests between French and British colonial powers to control the territory, and the oldest villages in the area were described by informants as lying "*bisa tangaraho*" (which means "on the telegraph line" in Hausa, perhaps referring to the telegraph poles that were used to demarcate the boundary between the British and French territories in the 1890s). Even so, none of the village elders could confirm that the old colonial border between French and British military territories passed through Guidan Wari or that this was why the original settlers located there. It remains only an intriguing possibility.

Nearby villages that were founded in these early years of colonial rule include Allah Karebo, which was founded between 1905 and 1915 by settlers from Alhassane to the south. Mallameye and other villages southeast of Guidan Wari were founded between 1930 and 1933 by Islamic scholars who came from the old city of Tessaoua in the south. Founded about 1930, the village of El Sangia is very close to Guidan Wari, only a little more than one kilometer west of the market, over the *arrondissement* boundary in Dakoro. People were living in El Sangia more than sixty years ago, but the village in the administrative sense did not exist yet. Guidan Antou was also founded around 1930 by villagers from Guidan Baguari south in the seasonal riverbed. Dadin Tamro was founded in 1933 by Bougajé from Guidan Wari. Dadin Tamro used to be a *quartier* of Guidan Wari, but the villagers petitioned to incorporate their own village administratively. Guidan Boujia was founded between 1935 and 1940 by people who came from Dan Malam. Kollo, to the south of Guidan Wari, was founded between 1940 and 1945 by the father of the present Kollo chief. In all, more than half of the villages in Guidan Wari *terroir* were founded between 1890 and 1960.

These villages that formed during colonial rule did so because the French had secured the area and made settlement possible outside the fortified walls of established towns. In addition, the climate was beneficent throughout the 1930s, and markets for groundnuts allowed many villages, even those established later without good road access or their own wells, to enjoy a degree of prosperity. There were no major famines during this time, and there was a significant expansion of agricultural land that continued for decades, until after Niger gained its independence in 1960.

The French had planned to build an all-weather road between Tessaoua and Dakoro that would have amounted to an improvement of its *route désertique*, but the road was never started. If the construction had occurred, Guidan Wari might have had the function and importance of an *arrondissement* capital. At some unrecorded and unremembered point in time, the French closed their garrison and left.

After independence, settlement was marked primarily by new village formation in the form of "cell division," in which villages formed by splitting off from other villages. After 1960, there was an order by the *sarki* (king), most likely the Mayahi *sous-préfet*, to gather people and form a town.[13] The Bougajé people were summoned from their camps about two kilometers to the northeast, and they moved into the southern part of the town. The Hausa people were already living near the market and in the northern part of town. In a 1955 aerial photograph of Guidan Wari, the market sits at the hub of all the walking paths in and out of the village, but there is hardly a village there.

Settlement, land use, and the local economy were all severely tried by a series of droughts starting in 1968. The drought known in the region as *Dan Koussou*[14] that occurred between 1968 and 1974 served as a catalyst for change in Guidan Wari. Droughts and subsequent famines form a substantial part of the historical record and function as temporal and mnemonic mileposts. In addition to *Dan Koussou*, according to Guidan

Wari residents, noteworthy droughts in recent memory include *Bahari* of 1982–1984, *Jabali* of 1965, *Mai Kwakari* of 1944, and *Balange* of 1942.

Among the effects of *Dan Koussou* was the devastation of cash cropping, especially of groundnuts. The 1968–1974 drought killed groundnut culture in Maradi Department, with production falling from 136,000 tons in 1967 to 15,000 in 1975 (Grégoire 1992). During *Dan Koussou*, there was a heightened need for subsistence food and stocks to feed the exploding cities. In Guidan Wari, responses to the drought included the liquidation of livestock herds, the sales of household assets, and further conversion of *daji* (bush land) to crop agriculture. The extensification effectively obliterated the buffer of *daji* that had been dedicated to grazing livestock, scavenging wood, foraging for medicinal plants, and many other uses, which farmers formerly could use periodically to respond to variations in rainfall. The virtual disappearance of *daji* surrounding the settlements is a change mentioned frequently, after drought and the construction of the new road, by villagers in Guidan Wari. Throughout the drought period, new settlements formed through "cell division": Farmers set out to convert the remaining bush land. The last new villages to form in the *terroir* formed in the early 1970s, as the drought forced farmers to convert more *daji* to fields. Koudidi and Dan Malam, the villages to the northeast of Guidan Wari and the west of Dadin Tamro, are each about twenty to thirty years old, formed at the same time or just after the *Dan Koussou* drought. When I visited in May 1995 to confirm the settlement dates of Koudidi and Dan Malam, I found these villages to be virtually deserted.

These and countless other later-settled and sometimes nameless satellite villages were formed during a crisis time and never had the chance to thrive that older villages had. They are farther from wells and fields, and often they lie atop active sand dunes. When they were founded, all the good land was already taken, and all the wells were dug. Today these villages suffer from more problems of deforestation and erosion than the older villages. They are a main source of temporary, distressed population displacements. No other factor in these marginal places explains the locational specifics better than the date of settlement.

After the drought, the village of Guidan Wari and surrounding *terroir* recovered its abandoned routines and experienced another round of short-lived prosperity. In 1978, the government of Niger gave Guidan Wari village its "center," with the construction of a plaza for village gatherings, a new mosque, and a television set. These developments corresponded to the uranium years. A path leading from the market to the present-day laterite road and two cement wells that are in use today were constructed then. For tax purposes, the market area was moved by the *sous-préfet* one kilometer out of Dakoro *arrondissement* and into Mayahi.

In 1992, a new laterite road was built that connected Guidan Wari with Dan Maïro to the south. The villagers like the new road because there is less sand, and millet and other blessings can be brought in more easily. The market has grown as a result, and products like cowpeas, cassava, and calabashes are easier to sell. In terms of population growth, the village of Guidan Wari has changed comparatively little compared to the newer villages that surround it, since it "filled" its immediate surrounding area with fields earlier than the satellite villages. Only since the end of *Dan Koussou* has the surrounding area experienced the same degree of land conversion.

It is difficult to tell exactly where the population is growing now in Guidan Wari. Measurement problems prevent accurate counts of individuals within households. The presence or absence of population growth is not as relevant as the fact that the villages of the area have consumed nearly all of the land that their founders claimed when they began their treks northward and southward into heretofore unsettled territory earlier in this century. The availability of land has played a critical role in the ability of rural peasants, such as those in Guidan Wari, to survive droughts. Since the turn of the century, land has gradually been converted from bush to fields. As the land became occupied, the villages lost their year-round grazing land and their buffer for use during famines. Because of the critical role land availability has played in the food security of rural peasants, it demands comprehensive treatment. Chapter 3 will undertake analyses of land-use intensity changes in Mayahi *arrondissement* and in Guidan Wari *terroir* in order to isolate exactly where the changes brought on by land conversion have occurred.

This chapter has documented the process of settlement in the Sahelian zone of Mayahi, occurring through the interplay of climate, conflict, and colonialism, which are here the driving forces of settlement. Population growth can exist conceptually out of its social and environmental context, but it has far more meaning when examined geographically, where causes and consequences are acknowledged to be different in different places. Observations about settlement dates give a starting place to study when the access to free land changed and the frontier of the Sahelian zone began to close. As Figure 2.6 shows, the villages settled later do not necessarily suffer from worse land, but the lack of the bush land buffer played a larger role in the history of these villages. As we will also see when we explore the changes in land and environment that occurred in Guidan Wari *terroir*, access to abundant land after 1975 diminished considerably during and after the *Dan Koussou* drought, and the options available to sedentary farmers were so reduced that their recourse to seasonal and temporary displacement became inevitable.

The story of the peopling of the Sahelian zone is in some ways a universal one. In the frontier phase, the bush provided the margin to accommodate deficits and surpluses in rainfall. Precolonial mobility included short-distance movements for the purposes of herding, hunting, and the conquest and conversion of bush land to crop agriculture. After the bush became filled with fields, the villagers lost the flexibility they had once had to withstand climate fluctuations. To provide the needed margin, extralocal movements to tap into extra-agricultural opportunities were then used. This transition in mobility types stems from the sequencing and process with which the Sahelian zone was settled.

The dynamics of settlement and population should be apparent; in critique, local settlement dynamics should be viewed not solely in microscale isolation but in a larger regional context of urbanization. Most of the movements in rural Maradi today are short-distance rural-urban mobilities, directed toward the department capital or toward cities in northern Nigeria, with the objective of engaging in commerce or of seeking sporadic short-term employment. The transition from exploiting the frontier to employing circular seasonal mobility could not have occurred without the organizing capability of the city.

For regions of the world still recovering from periods of colonial domination, it is admittedly difficult to unravel the recent history of colonialism from a population's long-term state of vulnerability. Records were often not kept before the colonial era, and vulnerability is often hypothesized to have begun during the colonial encounter. Though precolonial documentary evidence can be scarce, it is only by delving into this period that one can identify whether ecologically unsustainable situations predated their observed consequences.

Notes

1. An isohyet is a line that divides a region into areas receiving the same range of rainfall.

2. Based on meteorological data from Direction de la Météorologie compiled by AGRHYMET, for the years 1967–1990.

3. Niger's population statistics conform to standard demographic models. With its high fertility and mortality rates, Niger falls into the category of a so-called pretransitional society. Although the crude birth rate was 56.8 per 1,000 in 1960 (Faulkingham and Thorbahn 1975) and 58 per 1,000 in 1992 (DHS 1992), the crude death rate dropped from 31.8 per 1,000 in 1960 to 23 per 1,000 in 1992. This imbalance has led in the aggregate to a condition of rapid population growth.

4. Bougajé are normally thought of as former slaves of the white Tuareg who ruled central Niger. According to Chief Illa Boukary, the formerly pastoral population of Guidan Wari is Absanaoua, or Janbanza, not Bougajé. They consider themselves to be black Tuareg, to be differentiated from "red" Tuareg, but they

claim never to have been slaves. They *had* slaves, Illa says, but they made the slaves their children. Intermarriage makes genealogies difficult to interpret.

5. In his study of the colonial division of Hausaland, William Miles comments, "On the most basic level, the fact that Niger is a poor country in comparison with Nigeria is itself a result of colonialism and, more specifically, the colonial partition. To a large extent, the material fates of today's independent Federal Republic of Nigeria and the Republic of Niger were sealed at the turn of the century. When the British managed to claim for themselves a heavily populated, resource-rich, and topographically diverse territory and the French were left with a harsh, vast, underpopulated, largely Saharan expanse, the economic prospects of Nigeria and Niger were already set in place" (Miles 1994, p. 172).

6. Apparently numbers were boosted by the return of Tuareg refugees from northern Nigeria and the transfer of some territory from Upper Volta to Niger (Fuglestad 1983).

7. Land-use intensity is the percentage of total land used for cropping (Harris and Bache 1995).

8. Data on soil and vegetation came from field-based investigations; land-use and intensity data were based on the interpretation of aerial photographs; and climatic zonation data were based on precipitation data collected for 1977 to 1982 as part of the original 600-village study. One village in each of the zones was used for more in-depth study. The Projet Maradi agro-ecological zonation project covered the entire Department of Maradi. I made use of the results for Mayahi *arrondissement* only. A detailed description of the zonation project will be presented in Chapter 3.

9. It should be noted that while categorizing growth rates by ecological zone sheds light on some possible environmental drivers of population growth, it does not explain all variation. Within the agro-ecological zones there is hidden heterogeneity: Some villages within the zones experiencing negative growth did grow.

10. The zone with the fastest population growth, B_2, was unfortunately not included in the video transect.

11. For instance, taxes are collected by the *chef du village*, then given to the *chef du canton*, and then collected by the *sous-préfet*. This money goes to pay the teachers and nurses who work in the village, and any surplus is returned to the *chef du canton* and *chef du village*. In the case of disputes, the *chef du village* is consulted first, but if the instance is very serious or requires legal action, the *chef du canton* or the *sous-préfet* is asked to intervene. What is significant is the self-contained nature of village governance.

12. The 1995 estimate of 3,500 is based upon 1988 census figures and a uniform 3.2 percent growth rate. The numbers from 1988 are as follows: Guidan Wari, 775; Guidan Wani, 363; Kollo, 299; Guidan Bouzou, 159; Dadin Tamro, 279; Guidan Boujia, 209; Mallameye, 305; and El Sangia, 300. The five-kilometer circle contains 78.5 square kilometers of land.

13. There was some disagreement among the informants about when this actually happened. Some said fifty years ago; others said fifteen. It probably occurred between 1960 and 1965.

14. *Dan Koussou* means "son of a rat" in Hausa.

3

Land-Use Change Along the Environmental Margin

The Sahel is often characterized as a region in ecological decline. It has been drying out over millennia, and this process has accelerated recently because of the effects of human settlement. The confusion and disagreement over causes—especially whether people or nature is more to blame—is compounded by a plethora of political stances. Recent environmental changes are more easily attributable to human actions, though the distant past cannot be judged for such violations. Despite the scarcity of human historical evidence, the Sahel has been occupied or walked across for thousands of years. In this deep historical past, when a wetter climate permitted permanent agriculture in areas that are now well into the Sahara desert, earth and human history are mingled with uncertainties about what villagers and researchers alike say has "always" occurred.

A case was presented in the last chapter for considering the *process* of frontier settlement as it affects a region's growth and development and, by extension, its vulnerability to famine. The Sahelian agro-pastoral system combines small place-based investments with human mobility—including hunting, herding, and long-distance trade movements—to adapt to spatial and temporal variability in rainfall distribution. Abundant land is required to allow the land to recover nutrients recovery, to graze livestock, and to sequester famine foods. Such conditions existed in Maradi north of the fourteenth parallel from the time of settlement in the late nineteenth century to approximately the 1960s, and later in the more northern zones. As the Hausa production system moved north to colonize marginal land, the conditions for the growth of agriculture became increasingly precarious. This chapter will document in more detail the recent changes in land-use intensity in Mayahi *arrondissement*, in the Sahelian northern part of Maradi Department, looking specifically at the role agricultural land availability plays in food production capability, to show that changes in land-use intensity have eroded the ability of rural villages to use a bush margin to produce sufficient grain. This has created food deficit stress and increased the likelihood of population displacements.

When the persistent subtropical high pressure cells are in place over the Sahara Desert in the winter months, the Harmattan brings dry, dust-laden wind from the northeast. This dry season lasts from late October through May. No or very little rain falls during these months. The summer monsoon starts between April and June and moves in from the south, bringing water-bearing southerly or southwesterly winds that originate over the Atlantic and Indian Oceans (Hulme and Kelly 1993). Summer monsoons marking the annual northward movement of the intertropical convergence zone (ITCZ) deliver pounding rainstorms that often result in high levels of flooding and produce seasonal rivers and ephemeral lakes.

Prior to this, during the April and May hot season, humidity drifts up into the Sahelian zone from the south, a sign that the undulating ribbon of tropical moisture has reached northern Nigeria and begun to deliver the annual rains there. With the moisture in the air, days are hazy and almost unbearably hot. Anticipation fills the air; everyone waits for rain. Thunderheads propelled by tropical easterlies along squall lines running southeast to northwest give a promise of precipitation. But for many weeks, the only hints of the imminent rainy season are the humidity and haze.

When the torrential rain finally appears in June, it is preceded by a wall of windblown dust that sometimes blocks the sun entirely for a half hour or more. The scattering of concentrated storm cells—which can make weather conditions vary tremendously over one or two kilometers—makes rainfall accumulation very difficult to predict or indeed to measure. Although drought risk increases as annual precipitation means decrease northwards, the variation in rainfall at the local scale further reduces the assurance of adequacy. After a sizable rain hits Guidan Wari, villagers dig their hands straight down into the soil to see how deeply the water has penetrated. If the moisture envelops their fingers and touches their palms, then the ground is wet enough to plant in, and virtually all able people head to their fields to sow seeds. (See Photos 3.1–3.5.)

As this chapter will document in detail, much of the available land in Maradi Department now—even in parts of the northern Sahelian zone—is planted in crops. Figure 3.1 maps the changes in areal extent (percent of total area) from 1988 to 1995 for four major crops (millet, sorghum, groundnuts, and cowpeas) by *arrondissement*, for Maradi Department, based on official agricultural data from the Government of Niger (Direction de l'Agriculture 1996). The official statistics show that agricultural production has kept pace with population growth through increases in the area cultivated rather than through improvements in yield.

Photo 3.1 *Against the backdrop of an approaching storm, Guidan Wari village is not a self-contained world unto itself but one that relies on connections with the outside to maintain its survival.*

Photo 3.2 *Household members sow seeds of millet after the first rains have fallen. Conversion of land from bush lands to fields has proceeded to the point where the territory is almost entirely crops.*

Photo 3.3 Women winnow pounded grain in the shade. Converting millet into preserved edible food requires several labor-intensive steps.

Photo 3.4 Millet is both a subsistence grain and a cash crop in better years. Here men display the fruits of their families' work at the local market.

Photo 3.5   Millet is stored on the husk in granaries such as these. Protected against moisture and pests, it can stay edible for one to two years.

In a region where precipitation variability makes crop agriculture too risky to be the sole guarantor of household food sufficiency, the current rate of extensification is worrisome. Historically the region has always produced items for trade and has been food sufficient only in wetter years; however, the region's food supply is still subsistence-based and depends heavily on the efforts of small landholders to maintain adequate production levels. Traditional farming systems in the Sahelian zone have been based on adaptation to fluctuations in precipitation. The expansion of the area under cultivation and the decline of soil fertility in areas where fallow periods have become shorter or have disappeared pose challenges even to simple maintenance of current production. At the same time, the geographical specificity of these changes in land use and the corresponding changes in settlement and population dynamics require us to examine the conditions in more detail.

The Conditions of Agricultural Growth and Population Mobility

Despite the dangers posed by attributing changes in migration patterns to natural increase in human numbers alone, there is something undeni-

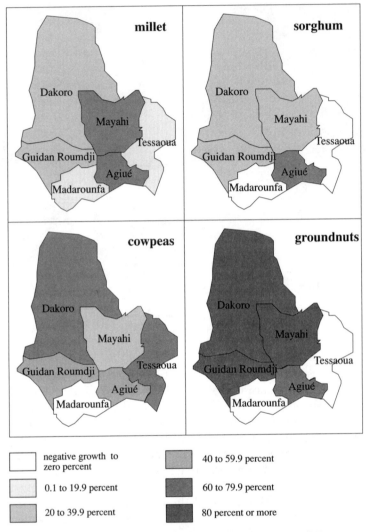

FIGURE 3.1   *Changes in Areal Extent for Four Crops, Maradi Department, by* Arrondissement, 1988–1995

ably arresting about the fact that the population of the Sahelian zone of Maradi Department grew a hundredfold in less than a century. Even if population growth is seen as a logical consequence of conflict, climate, and colonialism, its influence on agricultural production should be clear. It is the *periodicity* of the growth, that is, its tendency to occur in stages under specific historical circumstances, that highlights the role played by

TABLE 3.1 The Boserupian Transition in Land-Use Types

Stage in Transition Model	Length of Time between Crops	Population Density Range (persons per square kilometer)
Forest-fallow	20–25 years	0–4
Bush-fallow	6–10 years	4–16
Short-fallow	1–2 years	16–64
Annual cropping	several months	64–256
Multi-cropping	2 or more crops per year	256–512

Sources: Ester Boserup, *The Conditions of Agricultural Growth* (Chicago: Aldine, 1965), and *Population and Technological Change* (Chicago: University of Chicago Press, 1981).

mobility. Only through the analysis of population dynamics on the smaller scale—that of the settlement—can the processes of production and reproduction be linked.

By observing the relationship between growing populations and transformed agricultural systems, we can identify some of the general means by which farm productivity has been raised to meet the increased demand for food. As the possibilities afforded by extensification run their course and land becomes scarce, farmers are induced to achieve greater production by using more labor and more technical inputs, including irrigation and drainage in some areas and scientifically bred high-yield crop varieties that respond to chemical fertilizer. This process, in which agricultural innovation, induced by population growth itself, stimulates increased farm productivity, is most closely associated with the theories of Ester Boserup (1965). Boserup identified a developmental process in which land-extensive shifting cultivation systems become land-intensive sedentary cultivation systems, a process that she associated with population growth.[1]

Boserup has noted the positive correlation between population density and the intensity of land use. Societies living at low population densities tend to practice extensive agriculture, in which farmers cultivate a plot of land with a few simple tools and then move on to work another patch, letting the original plot lie fallow for perhaps as long as twenty to twenty-five years so its productivity can recover. Then the rejuvenated plot is cultivated again. Societies living at comparatively high population densities are much more likely to cultivate the same plot of land each year (that is, to practice annual cropping). These communities tend to have more sophisticated agricultural systems and farm tools. Levels of production per person are often higher under these more-intensive farming regimes, particularly if greater inputs of agricultural labor per person are required, in terms of time spent in cultivation. Table 3.1 presents the transitions in land-use types as specified by the Boserupian model.

There are some obvious limitations to Boserup's approach for marginal African agricultural systems.[2] Applied to the Sahelian conditions of the study area, Boserup's theory helps identify the constraints that inhibit the transition to higher production *despite* a persistent growth in population. The prevailing agricultural technologies derive from the cultural history of the in-migrating inhabitants of the region. Hausa settlers of Maradi in the nineteenth century were mainly from states in the wetter south and used techniques that originated and were mastered there.

The notion of constraints to agricultural transition has been pursued by agricultural economists following Boserup. Uma Lele and Steven Stone (1989) argue that the induced-intensification process in sub-Saharan Africa cannot occur autonomously; instead, they argue, it requires an active public policy to support broad-based agricultural development. Considering the combination of fragile African soils, declining rainfall, and contemporary, historically unprecedented population growth rates, rapid gains in productivity will be difficult to achieve. Lele and Stone suggest that in sub-Saharan Africa there are factors that limit the intensification predicted by the Boserupian model; these factors include substantial population concentration even in land-abundant countries, the tendency to "mine" the land for immediate survival, and potential expropriation of land from smallholders as value increases. In short, Lele's and Stone's prognosis is much less positive than Boserup's.

Other proponents of Boserup's theory use notions of mobility and access that are intrinsic to the land-use system. Prabhu Pingali and H. P. Binswanger (1991) utilize a sequenced model of settlement to illustrate the constraints to agricultural change in a heterogeneous environment. The environment includes "area X," which has loamy soils with high potential; "area Y," which has heavy waterlogged soils; and "area Z," which has shallow, sandy soils. A farming community initially settles in area X. The population increases, creating a predicament in which not enough food can be grown in area X using prevailing technologies. So the community starts to work area Y, draining, irrigating, and otherwise putting in labor to obtain a sufficient harvest of rice and therefore allowing the amount of food per person overall to increase. Eventually the population in area Y also increases to the point where food production per person decreases. Some members of the community then migrate to area Z, where the soils are shallow and light, easily eroded, and low in crop nutrients. Pingali's and Binswanger's point is that assuming continuous growth of population in a varied setting, especially on land that is incapable in some places of supporting intensive agriculture, creates geographically diverse outcomes. Clifford Geertz's research on agricultural involution in Indonesia (Geertz 1963) substantiates this. Boserupian processes do work, but often in fits and starts. One observation that emerges from linking settlement history and food sufficiency is that land with a lower potential can support its population for less time before it becomes necessary to move to another area.

Mobility therefore has potentially complex and significant relations to land use, with land-use change acting as both a cause and a consequence of mobility. Mobility is recognized as a barrier to the success of land-use management projects. In recognition of the constraints on successful transitions to sustainable permanent cultivation systems in marginal environments, however, Pingali lists government policies that prevent migration from marginal environments *despite* rapidly declining productivity levels (Pingali 1990). In Niger, the program of *gestion des terroirs villageois* (local resource management), along with the implementation of the resource tenure code in Niger, could very well fall into this category.

Agro-Ecological Zonation and Land Use

Compared with much of the rest of Niger, Maradi Department is relatively well endowed with workable soil suitable for agriculture. Much of this soil comes from the dunes that have covered much of the region since the start of the Neolithic period, perhaps 12,000 years before the present (B.P.), when the last major advance of the dunes was recorded (Butzer 1984). Brasset, Koechlin, and Raynaut (1984) divide Maradi's principal soil types into two groups: those of the ancient Paleolithic dunes, with a covering of red sand of aeolian (wind-borne) origin, and more recent Neolithic dunes. Paleolithic dunes have been modified by the wind and have formed in the shape of blurred rises that are oriented east to west and have a reticulated surface. In the interdune depressions, the vegetation grows thickly. The more recent dunes dating from the Neolithic period (around 8,000 B.P.) are also oriented east to west, but they have more pronounced peaks and steeper slopes and are shorter than the ancient dunes.

The dunes occupy the landscape of the northern and eastern parts of Maradi Department and make up the *jigawa* (upland) indigenous soil category. These soils are very sandy, low in organic matter, acidic, light in structure, and somewhat low in water retention capacity, though they can stay moist for two to four weeks in the rainy season, thus allowing germinating grains to remain alive until the next rainfall. Valley soils, including hydromorphic and isohumid soils of the arid brown group, cover a relatively small area. These soils contain large particles of sand, gravel, and clay, and they are too heavy to be worked easily. They also tend to be easily eroded by wind and water. Overall, despite a considerable amount of aeolian mineral deposition, the weak fertility of the soil, as well as the deficiency in phosphorus and nitrogen, requires fertilizer or long fallows. In many places, a hardpan of aluminum or iron laterite near the surface makes the layer of cultivable soil very thin (Brasset, Koechlin, and Raynaut 1984).

The researchers from Bordeaux participating in Projet Maradi undertook an agro-ecological zonation project for the department. They collected data on soil, topography, vegetation, precipitation, and land-use intensities and types and categorized them into six major groups and twenty-four subgroups.[3] Data on soil and vegetation came from field investigations, land-use intensity was measured based on the interpretation of aerial photographs, and climatic zonation was based on precipitation data collected for 1977 to 1982 as a part of the original 600-village study. (Refer to Figure 2.5 for a description of land-use types for the region.)

Most of the population is concentrated in the wetter southern zones (E_0, D_0, D_1), which also were the areas that were first settled. As Chapter 2 documented, population growth was highest in the wetter and more densely settled agro-pastoral zones (E_0, B_2, D_1, and D_0), and the lowest rates of growth were found in the northernmost and least-populated zones (B_0, C_0, and C_1). As the analysis here will show, these earlier and more densely settled zones had already experienced the closure of their agricultural frontiers by the time the French made their 1975 aerial photographic missions. What gives the closure of the frontier life is the process of land conversion on the local scale.

The "Received" Landscape

Throughout the Sahelian zone at the time of settlement, the terrain contained woody scrub growing on otherwise relatively barren plains and microenvironments of vegetation in swales between dunes. Trees included a small tree known as *katakara* in Hausa (*Combretum glutinosum*), a full-sized tree known as *kiryia* (*Prosopis africana*), and a shrub called *gwandar dawa* (*Annona senegalensis*) (Brasset, Koechlin, and Raynaut 1984; von Maydell 1990). This region has probably not been heavily forested since prior to 12,000–15,000 B.P., when the climate was wetter, but it was far greener in the past, with the scrub vegetation typical of today's Guinean zone. Grasses in the Sahelian zone were mainly annuals, with tall yellow grass serving as an indicator of long-fallow areas.

Illa Boukary, the elderly chief emeritus of Guidan Wari, remembers the vegetation being lush when he settled in the village around sixty-five years ago. "There were so many trees that you couldn't see your way—it was dark all of the time—you couldn't move for all the thorns and brush," he recalls. During the 1920s and 1930s, years after the area was first settled but before it became filled, trees included one called *anza* in

Hausa (*Boscia senegalensis*), an evergreen shrub up to four meters high that provided wood for fuel and poles for houses and fences. He also recalls *geza* (*Combretem micranthum*), a shrub also up to four meters tall; *sabara* (*Guiera senegalensis*), a shrub up to three meters high that is used for fuel; and *gao* (*Acacia albida*), which occasionally grew to twenty-five meters high. Called the "miracle tree" of Sahel, the *gao*'s wood is suitable for making tools, its seeds can be eaten during famines, and it sheds foliage during the rainy season, making it a suitable field tree (von Maydell 1990). Asked what happened to the trees and bushes that used to cover the area, Illa replied that "they were used in the house," mainly burned for fuel. Most of the trees were cleared after Illa arrived, while he was still young. He recalled that as a youth, there was sand on the ground, not grass.

According to the elderly informants of Guidan Wari, animals at the time of early settlement included *zabi zomo* (rabbit), *korege* (squirrel), *muzulu* (cat), *kura* (hyena), *tahimoussa* (bush cat), *bariou* or *maikuma koya mayna* (antelope), *dan jimini* ("son of an ostrich"), and *karembiki* (fox). These animals provided ample game for off-season sport but could not support a large and growing village population.

In the Hausa cosmology, bush land is regarded as the counterpoint to human civilization, the *dokar daji* (uninhabited wasteland) of nature, and home to animal spirits and ghosts. But even at the turn of the century when Guidan Wari was settled, most of the lions, elephants, giraffes, and other large game had already been hunted or chased out. Asked if he was afraid of the animals when he walked in the bush as a youth, Illa Boukary replied that he thought the animals were afraid of him.

This "received" landscape was slowly domesticated over the years and gradually converted to human uses for food, medicine, and fuel. To the Hausa settlers, the processes of field delimitation and landscape domestication were guided by the ancient animist rituals of land clearance. The act of clearing land could not by itself establish dominion over the space or protect the space from the evil forces that were believed to haunt the bush. To situate favorable forces on the land, especially in virgin spaces to be converted, a farmer had to appease a dangerous family of divinities, the clan of *mahalba* (hunters) (Nicolas 1966). A man desiring to chop down a tree had first to "tie up," or appropriate, the bush. Chewing a strand of *gamba* (tall grass), used to delimit fields and make fencing, and facing the tree about to be cut, he spat out the grass mixed with saliva in the direction of *gabas* (east), *gusum* (south), *yamma* (west), and *arewa* (north). A prominent characteristic of clearance is the precise orientation of the field in relation to the four cardinal points. *Gona* (fields) are generally square or rectangular and are thought to have a sexual polarity, with north and west being feminine and east and south being masculine. If the settler neglects to "tie up the bush" prior to cutting a tree, he risks having his hatchet bounce back and hurt him or having the tree fall on him and kill him (Nicolas 1966).

Thus tiny pockets of cultivated land, separated by uncultivated bush, tall grass, trees, and wildlife, gradually began to grow and spread apart until they hit other spreading, growing pockets of cultivated land. This process has been replayed all across the Sahel. SOS Sahel's Oral History Project (Cross and Barker 1991; Slim and Thompson 1995; see also S. Thomas 1992) reports results from fieldwork done in villages across the Sahel that mirror the changes Illa Boukary described. In a decade or two, most of this invaluable repository of environmental knowledge and oral history will be gone, since apart from a few efforts to record it, it is almost never written down. Nigel Cross salutes this oral tradition when he writes,

> Satellite imagery and oral testimony both have a part to play in shaping our understanding of environmental change. The people on the ground, several miles below the camera, know exactly what has happened in their locality; they may be amused or interested by the broader picture, but they can't see that it tells them anything they don't already know. They have been taking mental snapshots all their lives. (Cross and Barker 1991, p. 126)

Environmental Change and Variability

Debate about whether changes in rainfall are natural long-term variation, human-caused short-term disruptions, or a combination of the two also mark the scientific literature on Sahelian environmental change. The authors of the 1983 National Research Council (NRC) report on the Sahel expressed caution over attributing environmental change to human actions (NRC 1983). Based on evidence in stream deposits, fossil dunes, and lake beds and supported by evidence from fossil pollen and lake microorganisms, the record suggests that there were two protracted episodes of wind-borne sand deposition between 7.5 and 5.5 million years ago, setting the establishment of the Sahara Desert at the onset of the Quaternary period, about 2 million years ago. The record also shows a very dry climate 20,000–12,500 B.P., at which time the last major advance of dunes and sandfields was recorded. Southward advances in dunes since about 7,000 years ago have been comparatively limited in extent (Butzer 1983). At a smaller scale, the climate of the Sahel has been relatively constant during the past 2,500 years, with short- to medium-term oscillations between drier or more humid conditions and with rapid and highly variable climatic change occurring within these more predictable fluctuations.

During the period between 7,500 and 4,000 B.P., the region where the Sahara is presently located began to receive substantially more rainfall than it does at present. Large shallow lakes formed, particularly in the southern half,[4] and lush grassland scrub vegetation extended over most of the area. This environment presented considerable potential for hunter-gatherers, fishers, and herders. Within what is now the southeastern portion of the Sahara there developed a cultural pattern highly adapted to fishing and the harvesting of other forms of aquatic animals. The central Sahara was humid in the Neolithic period, with elephants, hippos, and crocodiles depicted in central Saharan rock paintings.

With a return to more arid conditions, which began by 4,000 B.P. and which were comparable to present-day conditions, food-producing people of the Sahara were forced southward toward the margins of the desert. It is in the Sahara Desert at this time that the first evidence of food production in Africa is found (Munson 1986). Bulrush millet was domesticated around 3,000 B.P., making the Sahel belt one of the agricultural hearths of Africa (Delehanty 1988). The climate was drying, and interior lakes and rivers began to disappear after 3,000 B.P. Resident populations had begun to permanently out-migrate east to the Nile, north to Berberland, and south to the Guinea coast. To these new locales, the migrants brought knowledge of savanna grains (Munson 1986). By 2,000 B.P., farming was common everywhere in West Africa *outside* the Sahara.

At a smaller and more recent scale, evidence for environmental change in the Sahel occurs in recorded history. Sharon Nicholson (in NRC 1983) uses the chronicles of medieval geographers and historians, journals of European travelers, the archaeological record, and other sources to document major droughts—that is, those persisting for more than one growing season—that occurred in the 1680s, mid-1700s, 1820s and 1830s, and the 1910s. Generally arid conditions ensued from 1790 to 1850, and minor droughts occurred in the 1640s, the 1710s, the 1810s, the beginning of the twentieth century, and the 1940s. Relatively wet periods occurred during the ninth to thirteenth centuries and the sixteenth to eighteenth centuries, from 1870 to 1895, and during the 1950s. Without conclusive evidence with a widespread distribution, such as tree rings or pollen, it is difficult to tell whether recent rainfall deficits are simply part of a larger pattern of natural variability, or if they are acyclical and result from human activities.

Short-term (from 1930 to 1960 to 1990) rainfall fluctuations in the Sahel have been the focus of much study since the Sahelian drought starting in the late 1960s. J. O. Adejuwon, E. E. Balogun, and S. A. Adejuwon (1990) articulate the need to collect more than monthly precipitation totals and, instead, to measure the length of the wet season and the onset, peak, and retreat of the rains in order to gauge the productivity of the rainfall for pastures and crops. M. V. K. Sivakumar analyzed rainfall records for Niger from 1945 to 1988, and he found pronounced contractions in the growing season across the country (Sivakumar 1991). For Maradi Department, the average date for the beginning of rains has moved from June 10 (1945–1964) to June 24 (1965–1988), the ending of the rains has moved up from September 29 to September 19 (for the same time periods), and the average length of the growing season has been reduced from 111 days in the 1945–1964 period to 87 days in the 1965–1988 period, a reduction of 21.6 percent. Sivakumar also notes that shortening growing seasons have been marked by an increased standard deviation of rainfall, meaning that cropping has become increasingly risky. Yet he concludes that although there is clear evidence that rainfall patterns were changing, there is no clear evidence to forecast either a long-term increase or a long-term decline in precipitation in Niger (Sivakumar 1991). Such variability may simply be endemic to the region rather than the result of purposive long-term anthropogenic change.

The disagreement about what causes the precipitation shortfalls have fueled debates on the effects of climate change and land-use practices. Compton Tucker at NASA's Goddard Space Flight Center tested the climate change hypothesis by analyzing the connections between changes in ocean circulation and drought in the Sahel. Tucker used an index of active vegetative cover, derived from the advanced very high resolution radiometer (AVHRR) data, to determine the extent of the Sahara Desert between 1980 and 1989. His analysis showed that the variations in the coverage of surface vegetation in dryland regions is governed to a large extent by interannual variations in rainfall (in Hulme and Kelly 1993; see also Tucker, Dregne, and Newcomb 1991).

At the level of the settlement, historical land-use change on the temporal scale of decades and in the spatial frame of the *terroir* (village group) reveals a pattern of interaction between variability and land use that requires closer inspection. Evidence of human occupation in the Sahel dates from 600,000 B.P., and since that time, selective hunting and gathering, bush fires, agriculture, herding, and other activities[5] have contributed greatly to the modification of Sahelian ecosystems (NRC 1983). No areas, however remote from human settlements, have been left undisturbed. Changes should be conceived as a series of reversible transformations that increasingly serve human ends. The progressive transformation of the landscape has meant an increasingly populated Sahelian fringe, supporting ever larger numbers of human inhabitants. The "eating" of the bush land made its use as a buffer in times of drought impossible. Land-clearing activities have led to reductions in biological diversity. Soils in the Sahelian zone, though workable, are generally incapable of supporting continuous uses. They easily become eroded and depleted by water and wind, and this is compounded by the removal of vegetation. If they are farmed continuously, the soils lose the nutrients essential for plant growth, a process known as soil mining (van der Pol 1994).

A study undertaken in Mali, which has comparable ecological and land-use conditions to southern Niger, found substantial reductions in soil nutrients. Forms of nutrient export (*from* the soil) include uptake by crops, leaching, erosion, and volatilization (passing off into vapor). Forms of import (*to* the soil) include crop residues, nitrogen fixation, weathering, airborne deposition, manure, and fertilizer. In sum, 47 percent of the nitrogen, 3 percent of the phosphorus, and 43 percent of the potassium used in crop production were depleted from the soil (van der Pol 1994). In circumstances where nutrient deficits are endemic, drought itself is not the main cause of the critical shortfalls. According to the International Crop Research Institute for the Semi-Arid Tropics (ICRISAT) in Niger, the main limiting factor in Nigérien agriculture in contemporary times is not the quantity of rain but the quality of soil (Ancey 1988). The result of permanent agriculture in this region of light and infertile soils is biotic overdepletion, where more nutrients are removed than replaced.

This condition of deteriorating soil productivity is perhaps the most precise description of the process known to the United Nations as desertification (UNEP 1995). The popular image of "deserts on the march" all over the world, permanently changing pastures and croplands into sand dunes, is misleading. Research done in the Sudan by the University of Lund from 1962 to 1984 found no evidence that the Sahara Desert was spreading outward from villages and water holes in the Sahel (reported in Stevens 1994).

To a Sahelian farmer, who cannot help having an anthropocentric perspective, it is not pertinent whether the observed conditions are considered desertification or whether they are natural or not. Changes in albedo (surface reflectivity) from vegetation removal is thought to influence rainfall (Bryson 1973; Charney 1975), but more pertinently from the view of a local inhabitant, such changes also degrade the local environment by lowering water tables, encouraging soil loss by wind and water erosion, raising local temperatures, and reducing the economic value of common-property resources such as grazing lands. The remote sensing perspective employed by international researchers has no way to approach accurately the impacts of land-cover changes on people's livelihoods. On the ground, land degradation and population pressure on resources describe the same process, which forms the core of Sahelian environmental change.[6]

In this section, I have argued that climate variability and change in the region are not recent phenomena, and this assessment is supported by researchers, notably the contributors to the authoritative NRC report. Since the beginnings of recorded environmental history, the Sahelian zone has seen dramatic changes in both its appearance and its biotic productivity. On the human historical scale, the argument for privileging anthropogenic forces over simple variability ultimately relies on the cumulative effects of human land uses. Many of the transformations of the Sahelian zone are damaging from a biological standpoint, but from the standpoint of a bush village, where the land must be made to work for human ends, the transformations are part of a bigger conversion of a marginal region. Much of the observed change occurs through a process of trading biotically abundant but economically unvalued plants and animals for those that can more directly contribute to a settlement's success as a human landscape. As evidence of human intentions, the present barren landscapes of the Sahelian zone are particularly striking reminders of this process.

Land-Use Intensity Change
on the Regional Scale

Given the variability of rainfall and variations in growing conditions in the Sahelian zone, we must examine land uses in Mayahi *arrondissement* at a finer scale. The precise dimensions of the changes between 1975 and 1995 in land-use intensity (LUI), or the percentage of the total land area that is covered by crops at a given time, were measured. The analysis was performed using remotely sensed land-use intensity data, specifically high-resolution aerial photos and digital videography.

Land-use intensity was initially measured in 1975 through a series of aerial photographs taken by the French military. From these, a topographic coverage for the entire southern portion of Niger was produced by the French mapping company IGN. The mosaic of 1:60,000 scale air photos was classified into three land-use intensities, using established criteria. The procedure for determining land-use intensity in the aerial photos was to identify initially the two extreme cases, leaving a third intermediate case. Thus a categorization scheme for zones that are complex but structurally homogeneous can be derived, creating the basis for the land-use intensity map of Maradi Department produced by the Projet Maradi team (Koechlin, Raynaut, and Stigliano 1980). This map was drawn to the same scale (1:200,000) as the topographic map series for Niger produced by IGN and shows fallow levels and soil and vegetation characteristics. Agricultural land use and fallow levels are depicted in three levels of intensity at roughly a one-square-kilometer resolution.

To measure changes in land-use intensity since 1975, a videography project was funded and implemented by USAID Niger (Brunner, Dalsted, and Arimi 1995). The advantages of videographed, remotely sensed land-use intensity data are ease of use and low cost. The data to account for twenty-year changes in land-use intensity came from a series of videotapes of global positioning system (GPS)-equipped flyovers of Niger, which produced transects with excellent resolution. Videography footage was converted in the GIS laboratory at AGRHYMET to digitized imagery, which can be viewed using generic Windows software. The program presents slightly overlapping images at a rate of one image per second, easily allowing the researcher to stop the "tape" and analyze individual images. To locate the images precisely, the camera was connected to a GPS unit that generates latitudes, longitudes and elevation and reads them into a spreadsheet. By matching up the frame number and the latitude and longitude, one can locate every single frame of the approximately 1,200-2,400-frame videography runs.

To compare the 1995 videographed land uses with the 1975 map of land-use intensities, a classification technique was used that was similar to that used in the Projet Maradi study. Initially, two extreme cases were delimited in the imagery: first, where the occupation of the soil is almost total and the uncultivated parts do not exceed 10 percent of the whole area; and second, where the cultivated parts do not cover more than 5 percent of the total land area. Between the two extreme cases, there exist patterns marked by small cultivated nuclei and fallowed fields in various stages. An application of the method of what the Bordeaux researchers call "convergent elimination" creates two further cases: first, where the cultivated fields are mixed with fallow but where clearing dominates and parcels in crops represent between 65 percent and 90 percent of the total area and second, the converse case, where the areas in fallow are more extensive than the parcels that are cultivated and parcels in crops represent between 35 percent and 65 percent of the overall area. In proceeding, one can detect the intermediate case where fields in crops and fallow fields are in comparable proportions, with the parcels in fallow representing between 35 percent and 65 percent of the overall area. The result is three land-use intensity classes (Koechlin et al. 1980).[7]

The results (see Figure 3.2) display a clear erosion of bush land throughout Mayahi *arrondissement*. Overall, out of 2,388 total points, there were 1,144 points in 1975 that were bush or fallow land (in which the LUI category 3), representing 48 percent of total points. By 1995, there were only 546 points, representing 23 percent of the total points. One-half or more of what was previously bush or fallow land in 1975 was by 1995 used for crops. From 1975 to 1995, about half of the bush land overall was converted to crop agriculture.

The erosion of bush land has occurred in both the eastern and western parts of Mayahi *arrondissement*. The western one was the more extensively farmed of the two in 1975, with 34.4 percent that was bush or fallow land and 46.5 percent that was mostly farmed. This means that almost half of the western transect was in the most-used category in 1975.

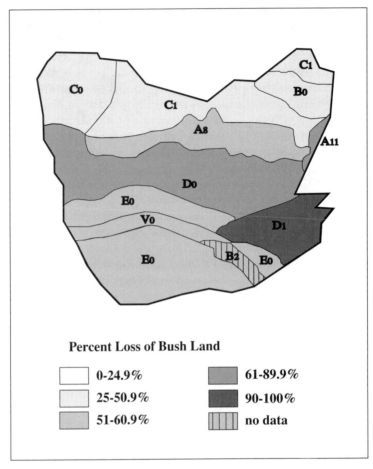

FIGURE 3.2   *Loss of Bush Land, Mayahi* Arrondissement, *1975–1995*

For the eastern transect, which is primarily a pastoral land-use type, 63.8 percent of the total land was bush or fallow in 1975, and 21.1 percent was mostly farmed. By 1995, 32.7 percent of the total land was classified as bush or fallow, meaning that 48.8 percent of this bush land was lost by 1995. In eastern Mayahi, therefore, about half of the land in the least-farmed category has been lost. Results are expressed in terms of changes in land-use intensity in Table 3.2.

The next set of tables itemizes the results further in order to determine more precisely how the land-use intensities have changed. Changes in land-use intensity are expressed in percent of total points in which land-use intensity stayed the same, increased, or decreased. Those displaying

TABLE 3.2   Land-Use Intensity Change, 1975–1995, Mayahi *Arrondissement*

Western Transect	Number of Points Total in Intersected Overlay = 1292				
LUI	*1975*	*%*	*1995*	*%*	*LUI change*
1 (65–100%)	600	46.5%	820	63.5%	+36.6%
2 (35–65%)	247	19.1%	284	22.0%	+15.0%
3 (0–35%)	445	34.4%	188	14.5%	–57.75%
	1292	100.0%	1292	100.0%	

Eastern Transect	Number of Points Total in Intersected Overlay = 1096				
LUI	*1975*	*%*	*1995*	*%*	*LUI change*
1 (65–100%)	231	21.1%	454	41.4%	+96.5%
2 (35–65%)	166	15.1%	284	25.9%	+71%
3 (0–35%)	699	63.8%	358	32.7%	–48.8%
	1096	100.0%	1096	100.0%	

stability amounted to 48.7 percent in the west and 41.3 percent in the east. Those that changed by being converted to fields amounted to 38.9 percent in the west and 47.6 percent in the east. Those whose land-use intensity changed from a LUI value of 1 to a LUI value of 2 or 3—from being abandoned, left fallow, or otherwise unused—amounted to 12.3 percent in the west and 11.0 percent in the east.

By 1995, 57.75 percent of the bush land from 1975 was lost. Of all the bush land in existence in western Mayahi in 1975 (with a total of 445 points), 113 points or 25.4 percent remained bush land, 192 points or 43.1 percent was converted completely to crops, and 140 points or 31.5 percent was only partially used for crops. Of all the bush land in existence in eastern Mayahi in 1975 (with a total of 699 points), 278 points or 39.8 percent remained bush land, 212 points or 30.3 percent was converted completely to crops, and 209 points or 29.9 percent was only partially used for crops. These changes are detailed in Table 3.3.

The breakdown of changes in land-use intensities by the Bordeaux team's agro-ecological zones (Koechlin et al. 1980) allows more inferences, especially about other areas that were not under the flight paths of the AGRHYMET video camera in November 1995. In these densely settled agro-ecological zones, bush land became very scarce between 1975 and 1995. In D_1 and E_0, the two wetter zones, bush or fallow represents less than 2 percent. In D_0, where Guidan Wari is located, bush or fallow land still represents almost 10 percent. Comparing the disappearance of bush land with intercensal (1977–1988) rates of population growth reveals the highest growth where land is most apparently scarce. In the E_0 zone, which was growing at an annual rate of 6.33 percent between 1977 and 1988, 61.1 percent of points did not change value. The land in this

TABLE 3.3   Land-use Intensity Change Broken into LUI Category

Western Mayahi changes, 1975–95

1975 LUI		1995 LUI	No. of points	% within group
1 (65–100%)	→	1	457	76.2%
n = 600	→	2	84	14.0%
	→	3	57	9.8%
2 (35–65%)	→	1	171	69.2%
n = 247	→	2	60	24.3%
	→	3	16	6.5%
3 (0–35%)	→	1	192	43.1%
n = 445	→	2	140	31.5%
	→	3	113	25.4%

Eastern Mayahi changes, 1975–95

1975 LUI		1995 LUI	No. of points	% within group
1 (65–100%)	→	1	141	61.0%
n = 231	→	2	41	17.8%
	→	3	49	21.2%
2 (35–65%)	→	1	101	61.0%
n = 166	→	2	34	20.4%
	→	3	31	18.6%
3 (0–35%)	→	1	212	30.3%
n = 699	→	2	209	29.9%
	→	3	278	39.8%

zone is already filled. Table 3.4 displays the results for the three most heavily settled areas.

Throughout the less densely settled agro-ecological zones, where density ranges from five to seventeen persons per square kilometer, there has been widespread conversion of bush land to crop agriculture. In northernmost zones of Mayahi, there were relatively few agricultural communities in 1975. By 1995, almost half of the land in the C_1 zone had a land-use intensity value of 1, indicating cultivation exceeding two-thirds of the total land, while 28.1 percent is still bush or fallow. Land that was abandoned, left fallow, or otherwise unused between 1975 and 1995 in these sparsely settled zones ranges from 0.0 to 16.1 percent. The conversion of the V_0 zone (Goulbin Kaba) should be noted, despite efforts to preserve it as a livestock corridor and to restrict farming there. Table 3.5 presents land-use intensity change for the lightly settled agro-ecological zones.

It is clear through the comparison of changes in land-use intensity that the variation by agro-ecological category is considerable, reflecting diverse conditions on the ground. Wetter land with less-compacted soil, such as

TABLE 3.4   Land-Use Intensity Change by Agro-Ecological Type: Densely Settled Zones (30–45 persons per square kilometer)

D_1 45 Persons/km² Growth Rate: 4.33%	1975 Points	% of Total in Each LUI	1995	% of Total in Each LUI	Change (%) by Category
1 (65–100%)	92	31.4%	227	77.5%	+146.7%
2 (35–65%)	91	31.1%	65	22.2%	28.6%
3 (0–35%)	110	37.5%	1	0.3%	–99.1%
Total	293	100.0%	293	100.0%	

E_0 33 Persons/km², Annual Growth Rate: 6.33%	1975 Points	% of Total in Each LUI	1995 Points	% of Total in Each LUI	Change (%) by Category
1 (65–100%)	288	68.7%	354	84.5%	+22.9%
2 (35–65%)	111	26.5%	57	13.6%	–48.6%
3 (0–35%)	20	4.8%	8	1.9%	–60%
Total	419	100.0%	419	100.0%	

D_0 31 Persons/km² Growth Rate: 2.35 %	1975 Points	% of Total in Each LUI	1995	% of Total in Each LUI	Change (%) by Category
1 (65–100%)	61	24.5%	135	54.2%	+121.3%
2 (35–65%)	103	41.4%	90	36.1%	–12.6%
3 (0–35%)	85	34.1%	24	9.6%	–71.8%
Total	249	100.0%	249	100.0%	

the E_0 zone, was often already filled by 1975. Results for this zone reflect this: 59.2 percent of the points with LUI values of 1 remained in that category. After 1975, land cover in the drier and more denuded areas of the north was increasingly converted to agriculture. Frames from the 1995 video footage reveal that even barren hills and gravel-covered seasonal riverbeds are being exploited by farmers for crop agriculture. Even so, land in the driest and most remote regions appeared to be still available.

These results show that during the period in question, bush and fallow land have clearly been consumed, particularly in less-favorable agro-ecological zones. In wetter and more productive zones, much of the available land had already been consumed by the time the 1975 air photos were taken. In the period between taking the air photos and the videography, available land in the more marginal A and B zones was eaten into, with some of the highest rates of change occurring in the areas with the least agro-ecological potential. In all the zones, 4.5 percent of the land reverted to lower land-use intensities through land abandonment, though much remained in the same category, especially in the favorable E_0 zone. These changes can be compared with population changes from 1977 to 1988 in the same agro-ecological zones.

TABLE 3.5   Land-Use Intensity Change by Agro-Ecological Type: Lightly Settled Zones (5–17 persons per square kilometer)

C_0 Agro-Pastoral Land Use Density: 17 Persons/km² Growth Rate: –1.85%	1975 Points	% of Total in Each LUI	1995 Points	% of Total in Each LUI	Change (%) by Category
1 (65–100%)	117	37.6%	119	38.3%	+1.7%
2 (35–65%)	15	4.8%	80	25.7%	+433.3%
3 (0–35%)	179	57.6%	112	36.0%	–37.4%
Total	311	100.0%	311	100.0%	

C_1 Agro-Pastoral Land Use, Density: 14 Persons/km² Growth Rate: –0.005%	1975 Points	% of Total in Each LUI	1995 Points	% of Total in Each LUI	Change (%) by Category
1 (65–100%)	186	31.2%	296	49.6%	+59.1%
2 (35–65%)	73	12.2%	133	22.3%	+82.2%
3 (0–35%)	338	56.6%	168	28.1%	–50.3%
Total	597	100.0%	597	100.0%	

A_8 Pastoral Land Use, Density: 11 Persons/km² Growth Rate: +1.42%	1975 Points	% of Total in Each LUI	1995 Points	% of Total in Each LUI	Change (%) by Category
1 (65–100%)	61	34.1%	55	30.7%	–9.8%
2 (35–65%)	20	11.2%	78	43.6%	+290%
3 (0–35%)	98	54.7%	46	25.7%	–53.1%
Total	179	100.0%	179	100.0%	

A_{11} Pastoral Land Use, Density: 6 Persons/km² Growth Rate: +1.24%	1975 Points	% of Total in Each LUI	1995 Points	% of Total in Each LUI	Change (%) by Category
1 (65–100%)	0	0.0%	19	32.8%	—
2 (35–65%)	0	0.0%	26	44.8%	—
3 (0–35%)	58	100.0%	13	22.4%	–77.6%
Total	58	100.0%	58	100.0%	

B_0 Pastoral Land Use, Density: 5 Persons/km² Growth Rate: –0.03%	1975 Points	% of Total in Each LUI	1995 Points	% of Total in Each LUI	Change (%) by Category
1 (65–100%)	8	4.0%	17	8.6%	+112.5%
2 (35–65%)	0	0.0%	39	19.7%	—
3 (0–35%)	190	96.0%	142	71.7%	–25.3%
Total	198	100.0%	198	100.0%	

V_0 Goulbin Kaba, Density: 12 Persons/km² Growth Rate: +4.46%	1975 Points	% of Total in Each LUI	1995 Points	% of Total in Each LUI	Change (%) by Category
1 (65–100%)	18	21.4%	52	61.9%	+188%
2 (35–65%)	0	0.0%	0	0.0%	0
3 (0–35%)	66	78.6%	32	38.1%	–51.5%
Total	84	100.0%	84	100.0%	

Despite the problem of comparing twenty years of land-use intensity change (1975–1995) with eleven years of population change (1977–1988), we can glean from Table 3.6—which shows population change and the loss of bush land by agro-ecological zone—a considerable heterogeneity in people to land ratios, with populations on better land having less of it to spare, and a large variation in demographic growth rates. Population has apparently grown especially fast in areas that have less land, such as the E_0 agro-ecological zone.

In this analysis of land-use intensity change, there are clear connections between the dynamics of settlement and population and those of land use. The rate of population growth is highest where land is scarcest, in the wetter zones with better soils. In sparsely populated land that is not suitable for agriculture, the populations are growing the least quickly, with negative population growth between 1977 and 1988 in three zones. From 1975 to 1995, half or more of the bush land in the more northerly zones was converted to agricultural fields. In other words, bush land conversion was occurring fastest in zones with the *lowest* population growth. Spatial variation in bush and fallow land, and especially differences between change in north and south in land converted to agriculture, reveal the role of land shortage in creating the stress of a closing frontier. There is likely to have been real population change throughout the *arrondissement*. In the north, the stagnant growth rates conceal what is likely to be considerable movements in and out. Land conversion in the absence of population growth is likely to be the result of farmers in established villages traveling farther to fields.

Seen in the light of Boserup's theory of population-induced agricultural change, these conditions of scarcity are prerequisites for the intensification practices. Land scarcity is diffusing across the *arrondissement*, following a historical progression: C_0 is in 1995 where D_1 was in 1975, which was where E_0 probably was in 1945 or 1950. But C_0 receives far less rainfall and suffers a far higher risk of drought. This natural progression is reaching its logical end, for there is nowhere else to find land. When all the remaining bush land is consumed, will villagers begin to put more labor into maintaining soil fertility and increasing yields? To answer this question in context, we must delve to the microscale level and examine the changes on the village level.

Land-Use Intensity Change
on the Local Scale

Of the agro-ecological zones from the 1975 classifications, Guidan Wari in the subgroup D_0 should be regarded as typical. This zone features dunes of

TABLE 3.6 Population Change and Loss of Bush Land

Agro-Ecol. Zone	Area (km²)	No. of Villages	1977 Total	1988 Total	Average Size	Pop Change 1977–1988 (annual)	% Loss of Bush Land
B_2	103.62	6	822	1779	297	7.02%	—
E_0	1557.42	113	25690	51563	456	6.33%	-60%
V_0	253.37	9	1936	3163	351	4.46%	-51.5%
D_1	498.33	38	14145	22767	599	4.33%	-99.1%
D_0	1478.09	99	35835	46413	469	2.35%	-71.8%
A_8	663.33	17	6562	7673	451	1.42%	-53.1%
A_{11}	29.16	1	158	181	181	1.24%	-77.6%
C_1	1059.58	43	15544	15536	361	-0.005%	-50.3%
B_0	261.68	4	1358	1354	339	-0.03%	-25.3%
C_0	619.49	44	13184	10762	245	-1.85%	-37.4%
Totals	6524.07	374	115244	161191	431.0	3.05%	52.3%

mild relief, sandy soils with favorable water-retention capability but weak fertility, and strong climatic constraints. The region surrounding Guidan Wari was in fact used by pastoralists for decades prior to the 1890s.

The process of frontier settlement continues to be guided by sociopolitical structures of inherited indigenous rule and customary land tenure. In the Sahel, land tenure rights are frequently allocated to the first settlers, and later their descendants, who occupy the territory of the village. Such "rights of the ax" are allocated to the heads of clans or of lineages that were involved in the original clearing of the forest for field crop production (Freudenberger and Mathieu 1993). Sometimes, too, the land tenure arrangements in Sahelian settlements are determined by the "rights of fire," whereby a brush fire is set and the new village occupies the area it burns. Throughout the sedentary parts of the Sahel, the chief has considerable power to allocate land and settle disputes over land, crops, trees, and wells. Chiefs have absolute authority over the territory of the village, but in cases involving bush land or land without clear authority, such as in the seasonal river valleys, a higher governmental authority must be consulted.

Control over land allocation in Guidan Wari is considered "customary." Land is not owned by anyone; rather, it is retained by families and controlled by the chief. The issue of who gets land is also customarily based. Fields are normally allocated to the head of household, then subdivided among the family into plots. Whether wives and other individual family members, including women, get their own fields varies by region. In Guidan Wari, women usually help with their husbands' fields and do not have their own plots, except in the case of a project-sponsored vegetable garden near one of the village's two wells. Otherwise there are no intensive gardens, except for a few fenced cassava plots.

According to the chief, land is given to any stranger who wants to move to the village. If no land is available, then the chief will rent some of his own land. Disputes over land in Guidan Wari are not rare, but normally they are resolved easily with the intervention of the chief, who consults the *chef de canton* or the *sous-préfet* if necessary. While I was staying in the village, I heard about a few grumblings related to grazing land and access to the village wells. In all, though, there is a surprising lack of conflict between farmers and pastoralists. Fulani herders pass through Guidan Wari frequently on their way from Goula to Dan Maïro. There were no reports of conflict. Flare-ups often do occur at wells, and many of these are apparently quickly resolved. The chief's explanation for the lack of conflict is based on ethnicity: The Bougajé and the Fulani regard each other as "brothers" and do not antagonize each other the way the Hausa and Fulani do.

Land and resource tenure in Guidan Wari enjoy stability because of the power and respect commanded by the chief. As of 1993, tenure was becoming a hotly contested issue in Niger, owing mainly to the advent of reform in resource tenure laws in the country. Such dynamics are indeed integral to, and representative of, the level of meddling in which the national government of Niger has engaged in an effort to maintain control of the country. Chapter 4 will look at such dynamics in more depth, in regard to the government's efforts to reduce levels of seasonal migration.

The population density of the area immediately surrounding the village of Guidan Wari is higher than that of the zone as a whole, with an estimated 44.5 persons per square kilometer, as compared to 31.4 for the zone overall. The land-use intensity change analysis calculated the proportion of land that was converted from the least intensely cultivated (category 3) to the most intensely cultivated (category 1) between 1975 and 1995 as being 14.5 percent of the total land. In the D_0 zone overall, conversion to farmland amounted to about 46 percent, with 47 percent staying the same and 6.4 percent reverting to bush.

In Guidan Wari, real land shortages are more acutely felt in the land immediately surrounding villages. Based on the analyses of the aerial photos and on field analysis, using the photos from 1955 as well as those from 1975, there has been a dramatic change in land-use intensity in Guidan Wari, from 47.7 percent in 1955 to 80.8 percent in 1975 to 96.8 percent in 1996. The aerial photos show gradual erosion of bush land in the vicinity of Guidan Wari.[8] This progressive infill emanates from the villages, eventually consuming all but a small parcel of land approximately four kilometers away (see Figure 3.3).

The implications of land-use intensity change are directly felt in the village. The sense of impending crisis appears to be influenced not just by the threat of drought and changing environmental conditions but also by the common perception that fields are inadequate to provide sufficient nutrition. I asked questions about the loss of bush land and of available land, about the use of manure and chemical fertilizers, and about the availability of oxcarts, firewood, and grazing land in nine villages in the vicinity of Guidan Wari. In a more general way, I sought to gauge villagers' level of awareness about the consequences of land conversion—whether they perceived it to be a problem. Some of the villagers' responses, which are based on group interviews conducted in the pre-rainy season, are recorded in Table 3.7, grouped by settlement date. The subject of adaptation will be covered more thoroughly in Chapter 7, when village-level consequences of circular mobility are discussed.

In the villages settled early, including Guidan Wari, Guidan Wani, and Allah Karebo, perceptions varied. Interviews in Guidan Wari and sur-

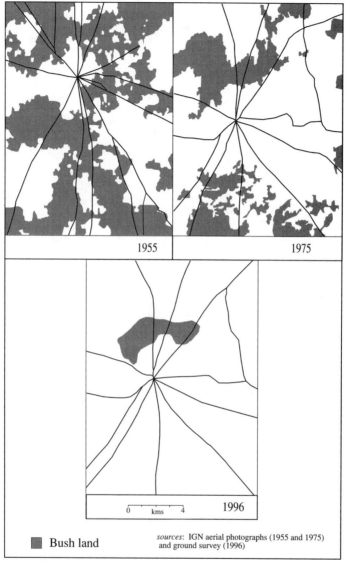

FIGURE 3.3   *Areal Extent of Bush Land in 1955, 1975, and 1996*

rounding bush villages found a clear connection between the closure of the agricultural frontier and the perception of their vulnerability to seasonal and annual shortfalls in food production. In Guidan Wani, located east of Guidan Wari, the level of awareness was very high. Asked if there was enough room for farming, informants laughed at how obvious the

TABLE 3.7 Village Land Use and Potential for Improving Production

Village Name	Founding Date (est.)	Pop. Estimate	Bush Land	Fallow Land	Oxcarts	Pasture	Wood Supply
Guidan Wari	1892	968	no	no	7	some near Tamro	sufficient for now but getting worse
Guidan Wani	1899	831	no	no	4	?	sufficient at the moment
Dadin Tamro	1940	500	no	no, not for 20+ yrs	0	a little	none (30–40 kms away)
El Sangia	1915	300 (Dakoro)	no	yes, in a few places	0	more cattle here than further east	?
Kollo	1945	600	no	no	2 (rented from people south of them)	yes, west of the village	not a concern
Guidan Boujia	1940	400	some	some do, some don't	1	bushland for animals	?
Allah Karebo	1930	150	some	some fallow fields	?	?	?
Mallameye	1935	846	no	no	0	?	?
Guidan Antou	1930	740	some	no	rent from people in area	a small area	?

question was and replied, *"Ba wuri, ba huta!* [No room, no rest]." There has been no bush land near the village for fifty years or more, and fields have been filled that long too. The village had four oxcarts for use in transporting manure, however. Allah Karebo was reported to have some fallow fields, and villagers did not use manure there. Yet in the interviews in surrounding villages, I was surprised how many farmers did not perceive any problem at all with fields filling the countryside. These individuals were not in the majority, but they were not rare either, and their perceptions were difficult to reconcile with the conditions they themselves were describing.

Among colonial-era villages, the level of perception also varied. In Dadin Tamro, located five kilometers to the northeast of Guidan Wari, the villagers reported that no fallowing was possible in the immediate vicinity and that the land had had no rest for more than twenty years. The villagers do use animal manure, which they carry on their heads to their fields. They know they are not resting their fields. There is grazing land to the north (with tall yellow grass but no trees) and to the west. The area to the south has not been farmed in a few years. In the bush area to the west, Gingé (an older but smaller village than Dadin Tamro) and Barouta (both in Dakoro *arrondissement*) are taking advantage of Dadin Tamro, eating into the remaining bush land. Dadin Tamro made a protest through the *sous-préfet* in Mayahi about this, and if nothing is done they will go to the *préfet* in Maradi. The villagers in Dadin Tamro reported that firewood was scarce and that they buy it sometimes or scavenge it from the bush, sometimes up to thirty or forty kilometers away. One person reported that in order to get any wood, "you have to go to Goula," which is located about seventy kilometers to the north.

Villagers interviewed in El Sangia, one and a half kilometers to the west of Guidan Wari, maintained that land is abundant, fallowing is routinely practiced, and manure is applied to fields "if they have time" (which provoked chuckles among those observing the interview). Farmers do use chemical fertilizer, which is bought in Maradi. There are also more cattle in El Sangia than in Guidan Wari and other villages to the east. I asked if other villages to the west, in Dakoro *arrondissement*, also had more free land. The response was yes, though the villagers also noted that some villages are eating into the remaining dedicated herding space. Even so, the villagers reported that there apparently seemed to be enough grazing land for the moment. They also said that no free land was available in El Sangia for crops, but the farmers do rest their fields, sometimes, perhaps because of their absences.

Other villages in the *terroir* report similar conditions. Informants in Mallameye reported no fallow fields, no extra fields, no manure, and no oxcarts. Guidan Antou, located five and a half kilometers to the south, has a small parcel of grazing land actively reserved for animals. There are no fallow fields, but villagers do put manure on the fields, depending on the availability of an oxcart. Villagers in Guidan Boujia, five kilometers to the east of Guidan Wani, reported that they purchase chemical fertilizer from Maradi and also use animal manure. They have one oxcart. According to informants in Dadin Tamro, the neighboring village, Guidan Boujia has more space in their area and their fields get a chance to rest. Guidan Boujia has also had better luck with rain.

In the villages settled later, the same levels of diversity in responses were observed. Villagers in Kollo, south of Guidan Wari, reported having pasture land near Tako Tako, but as with Guidan Wani, the fields have never rested since the villagers arrived fifty years ago. The villagers now use animal manure for their crops, which is transported by two oxcarts in the village, but they cannot afford the expense of chemical fertilizer. Wood did not seem to be in especially tight supply at that moment. According to the villagers, trees die and fall down in the fields, and nobody has to buy firewood. The informants agree that there are probably fewer trees now than when the village was settled, but there are more than during the droughts of twenty years ago.

These responses from the villages around Guidan Wari *terroir* reveal a generally high level of awareness that land is becoming scarce everywhere. Villagers understand that nutrients removed from the soil in the form of food were not being replaced. Given the comprehension of this need, I pictured migrants toting a sack of chemical fertilizer on their heads when returning home from their off-season travels. But land-use techniques in the Guidan Wari *terroir* are not, in fact, changing in response to the growing population. Chemical fertilizer was perceived to be too expensive. Overall, oxcarts were proposed as a solution to soil fertility problems far more often; so often, in fact, that I became rather skeptical of the responses. Emlyn and I both felt that we were being "steered" by villagers toward the subject of oxcarts, since oxcarts seemed to be both a needed item and something that I, as a "rich white visitor" and a friend of Emlyn, might be interested in providing them.

The area around Guidan Wari is typical of the Sahelian zone in that domesticated animals form a central part of the village's financial base. Sales, pledges, and rentals of animals effectively make the collective "stock market," with its individual owners, a local economy second only to that of grain. The fields in Guidan Wari are used as pasture for nine or ten months out of the year. During the rainy season, herds are sent north with sons or hired herders to graze in the Tarka Valley, about 100 kilometers away. When the herds come back to the area, they seem to know instinctively where home is. The rest of the year the animals graze wherever they can, and their manure enriches the soil.

In the villages one frequently hears the comment that the problem is that there are not enough animals; there just has not been the means available to build herds back up after the last drought. Animals serve as an insurance policy as well as a bank account. Were they substantial enough, the herds could enable villagers to make a sufficient surplus during good years to get out of the downward spiral where they have been stuck for the last twelve to fourteen years. Before *Dan Koussou,* they had more animals. Most of these survived the 1968–1974 drought but not the 1982–1984 drought.

Dependence on livestock has been criticized because of the problem of overgrazing. Based on the organic input needed to fertilize annual crops, Mike Speirs and Ole Olsen (1992) calculated that a pasture to field ratio of 26 to 1 was necessary to sustain an agro-pastoral system in this rainfall regime. This analysis, though, neglects the vital role that livestock plays, not only as a savings account but also in field fertilization. The role of manure is especially important when price supports for chemical fertilizer have been discontinued. Villagers are not likely anytime soon to allow their fields to go fallow and to invest their energies instead in herding. But were the capital available to invest in animals and to feed and maintain the animals in the household compounds instead of in pastures far away, there would be a significant amount of manure produced that could be applied to the fields. The problem is that one source of investment capital, remittances from the migrants who work in distant cities, is being used not for livestock but instead for the purchase of small quantities of rice and millet to help the residents survive periodic crises.

In a rural village setting, food sufficiency is a function of a household's ability to convert solar energy, received directly from the sun and indirectly through plant and animal protein, into something of nutritional value. In the Sahelian zone of previous centuries, where land had been used for nomadic livestock herding and sporadic shifting agriculture, drought was inevitable and endurable. Animals moved in search of pasture, and people followed them, leaving sometimes-degraded land behind to recover. Population densities were in the range of five or ten people per square kilometer, leaving plenty of room to move about. With the settling of the zone in this century and the rapid expansion of crop agriculture, the equilibrium was lost. Attempts by the Vichy French notwithstanding, the plow did not catch on here, but even so farmers with axes and hoes effectively cleared the land, and a domestication of the landscape took place. With the continued expansion of crop agriculture in 1950s and 1960s, the bush-eating frontier found itself with a problem, so Hausa and Bougajé villagers began to move again.

This chapter has examined the physical dimensions of land use that link settlement dynamics to demographic processes. I have narrated a

periodization of the process of frontier-conversion, as well as its consequences, looking in particular at the period from the end of the 1968–1974
drought, when livestock herds were low and farmers reacted to the
prospect of better rainfall by converting more land to crop agriculture, to
the present, when more and more villagers are allocating their time and
efforts elsewhere. During the period from 1975 to 1995, land-use intensities rose throughout Mayahi *arrondissement*, with the amount of land in
the category of highest land-use intensity roughly doubling and the
amount of land in the category of lowest land-use intensity being cut
roughly in half. Overall, 43 percent of the land covered by the analysis
experienced an increase in intensity, in the conversion of formerly bush
or fallow land to crops. Mapped by agro-ecological zone, the pattern of
land-use intensity change reveals similar conversion rates in zones that
are more heavily populated and more endowed with soil and water resources and in the sparsely populated and less resource-endowed zones.
Even the driest and most sparsely settled zones experienced very significant conversions of their bush land, even without any apparent net population growth.

Changes in land-use intensity have eroded the ability of rural villages
to use a bush margin to produce sufficient cereals, thereby creating food
deficit stress. At the local level, in the cluster of villages surrounding
Guidan Wari, group interviews revealed that villagers perceived land
shortages to be real. These shortages have exacerbated preexisting problems in the area. Reduced soil fertility, and not drought, is perceived to be
the root cause of the villagers' insufficient yields. Yet no changes were
being made to agricultural or land-use techniques in the Guidan Wari *terroir*, for reasons associated with drought risk and the lack of capital to invest in added inputs. The existing infrastructure is inadequate for the distribution of fertilizer to the fields where it is needed. None of this bodes
well for the future of Mayahi, or for Niger as a whole. Given the demographic structure of Niger, assuming no relevant changes in technology
and no new availability of additional arable land, Niger will have to continue to increase its agricultural production significantly under conditions that hardly favor permanent agriculture at all. The population of
Maradi Department is very likely to double again at least once before it
stabilizes.

Under only slightly different conditions, one can imagine an agro-pastoral economy that thrives in the Sahelian zone. There would be a more
dynamic market for food, fertilizer, and wood, more outlets for products,
and more livestock. In general, there would be more people and goods in
circulation and more active ties to the outside. Can seasonal migration
play a role in making this more of a reality? Given the area's apparent inability to feed everyone under given conditions and using existing tech-

nologies, in order for Guidan Wari to be viable under current conditions, a resource must be imported.

Notes

1. Boserup's theory was intended as a counterpoint to a Malthusian view, which stresses the determining role played by agricultural assets *upon* the local population, where a population's ability to feed itself on its existing resource base is strained by the numbers of mouths to feed. According to Boserup, the converse causal relationship of population's effect on agriculture is scarcely regarded. Boserup's thesis is that population growth acts as an independent variable determining the path of agricultural developments. From her cross-sectional perspective, she argued that population growth often actually worked as the major dynamic engine of agricultural change, stimulating the adoption of improvements in land use and technology. In such a causal relationship, no reciprocal effects are assumed.

2. One limitation is that treating the rising population density as an independent variable that itself is responsible for creating changed conditions does not account for what causes the rising population densities. In treating population growth as an abstraction, Boserup makes the same mistake as Malthus in attributing causal power to what is simply a yardstick. The second limitation is that the thesis of "conditions of agricultural growth" is rather short on actual conditions. There is very little explicit discussion of biophysical barriers to either expansion or intensification of areas cultivated. The third limitation is that the theory subjects local conditions to a stages of development model, which does not account for the social and political processes of agricultural change.

3. The project covered the entire Department of Maradi. I used the data for Mayahi *arrondissement* only.

4. Records for the level of Lake Chad, on the eastern border of Niger, reveal sharp periodicities, from high levels around 30,000 to 20,000 B.P., to a period of aridity around 20,000 to 13,000 B.P., to dry interludes between 12,500 and 9,200 B.P. A rapid rise of the lake to an intermediate level began about 9,200 B.P., lasting until 8,500 B.P. After the fall of the lake level to its modern position around 7,400–7,000 B.P., the Paleo-Chad maximum was achieved at +40 meters, fluctuating below this high level until 5,200–5,000 B.P. A last high lake level occurred between 3,500 and 3,000 B.P. Subsequent levels have fluctuated around those of the last millennium (Butzer 1983).

5. Bernus and Gritzner identify some agents of anthropogenic change, including: ecological simplification through bush fires; the trans-Saharan trade and its effect on wood cutting and charcoal making along caravan routes; agricultural expansion caused by the cessation of the slave trade (encouraging northward drift of agricultural production into formerly hostile regions); the proliferation of cattle, goats, and sheep; the development of modern transportation networks; and the explosive growth of Sahelian cities (Bernus and Gritzner, in NRC 1983).

6. UNEP's rebuff to antidesertification criticism reflects this: "Whether the deserts of the world are receding or not is not the issue. The process of desertification is not shifting sand dunes, or advance of a wall of sand, but rather patches

of increasingly unproductive land 'breaking out' and spreading over hundreds of square kilometers" (W. Franklin G. Cardy, quoted in Pearce 1992, p. 41).

7. The Bordeaux researchers reduced their 1:60,000 scale map to 1:200,000. Since the smallest polygon from the 1975 map was about one-half kilometer, I used this to generalize the videography data source. Although the 1975 map was produced for all of Maradi Department, I selected only the portion that corresponded to Mayahi *arrondissement*. This portion was scanned and digitized into the Arc/Info GIS program on a Sun Workstation platform. In all, 774 polygons representing the same three LUI classes were classified: an LUI value of 1 for land that is 65–100 percent in crops, an LUI value of 2 for land that is roughly half (35–65 percent) in crops, and an LUI value of 3 for land that is 0–35 percent in crops. Topology was built for these polygons in Arc. A few classification and mapping errors from the predigital age were encountered. The minimum polygon size for the 1975 map was 0.5 km².

The videographed transects were classified using the same three categories: The LUI value is 1 for land that is up to a third bush or fallow, 2 for land about half farmed and half bush or fallow, and 3 for land that is still mostly bush or fallow. The results were filtered according to the same criterion of 0.5 km² minimum polygon size and converted into digital coverages. The GPS unit used was a Trimble Pathfinder Basic Plus. This GPS model has no base station, so latitudes and longitudes should be regarded as somewhat approximate, with perhaps up to a quarter kilometer of error. The fixes were combined with the time-code generator and written to the audio track of the video camera in the airplane. Later the times and locations were converted to a text file, which could then be entered into the GIS. Digitization of the video image was performed to align with the 1-per-second GPS fix, on a 320 by 240 pixel (24-bit per pixel) video frame. The video images were stored as .avi files on a recordable CD-ROM and can be viewed with a generic video player on Windows. Problems encountered in this analysis included the imprecision of the GPS and a problem with the zoom lens on the video camera. Generalization criteria were based on a measurable feature from the video frame. I used the National Highway, which is conveniently exactly six meters wide and which the flyover crosses. From this known feature I could determine that the frame size for the Mayahi flights was roughly 100 by 140 meters. In a ratio scale, this translates to roughly 1:20,000. If the smallest polygon on the 1975 map was about 0.5 km², then this would be the equivalent to five video frames if the shorter vertical dimension were measured. So for classifying the video frames, any series of similar values that was smaller than five frames was dropped.

The text files containing the latitudes and longitudes of the two transects as well as the LUI values were entered into Arc/Info GIS as point coverages, and the changes between 1975 and 1995 were isolated using Arc/Info. The "point-in-polygon" intersect command in Arc slices the polygons of one coverage into points and overlays them atop the other point coverage, so that 1975 data can be compared with 1995 data and changes in the LUI values can be compared. East and west transects for Mayahi were kept separate in order to isolate spatial variation between primarily pastoral and primarily crop-agricultural land-use types. All the agro-ecological zones found in Mayahi except for one, B_2, were found to fall under the videographed swath. The western transect was settled primarily from the 1920s through the 1950s, whereas the eastern transect remained pastoral land until more recently.

8. The process of bush land consumption was well underway by the time the first series of photos was taken in 1955. Unfortunately, no photos in the archives are available to document the process of land conversion in this area prior to the 1950s.

4

From Caravan to Bush Taxi

Abou Hamza the sword maker left his village during the drought in 1971 and moved to Maradi. Today he embodies several common traits of successful seasonal migrants. The advantage of his livelihood is that it can be practiced anywhere, and he is almost guaranteed a market wherever he may choose to go. Abou sells his wares at the Maradi market and around town and earns a good living, working as much as he likes and selling everything he wants. Abou's skilled craftsmanship and his long residence in the city make his occupation virtually ideal. He has social ties in many places and he can easily pick up and move. The city rewards his skills with outlets for his products, and because the materials he needs are readily available, his skill is portable, making his livelihood well suited to local conditions. Most craft occupations require few resources other than human ones; necessities can be carried along. This in effect makes mobility practicable; for nonfarmers, there is no purely economic reason to stay settled. And because ties to the village group remain strong over distance, most who decide to go away briefly do not completely alienate themselves.

Why then is Western-sponsored economic development in this region fueled by the hopeful idea that reducing migration will somehow allow economic development to occur? If rural places were more "sticky," the presumed logic goes, migrants would stay home and make their communities more attractive to investment. The political economy of the region continues to emphasize alliance over accumulation, a trait that has a direct bearing on the practices of mobility. In the changes that have been observed over recent decades, particularly the transformation of rural livelihoods, the national government has played a noticeably weak role.

As in much of Africa, the household-based production, consumption, and income-generating activities in Niger are intermixed, making the economy and the environment very difficult to separate. One particularly salient aspect of the economy of the Sahelo-Sudanian zone, with its long dry season and overexploited resource base, is the function performed by nonagricultural income in satisfying food needs. The economic context of income generation spans a continuum of decisionmaking settings from the household to the region to society at large, and mobility plays a critical role within this economic context.

Economic Development in the Sahel

Like many countries in sub-Saharan Africa, Niger has experienced economic trends marked by extremely sluggish growth since the 1970s. Many countries among the former French colonies have seen a particularly persistent stagnation of their economies. Although membership in the CFA franc zone provides some benefits in the form of stability (the currency of francophone West Africa is tied directly to the French franc) and French technical and military assistance, the devaluation crisis in 1994 underscored just how fragile and dependent the economic system is. It also marked a turning point in fiscal oversight for the region—stalwart nationalists might argue a new low in neocolonialist meddling—in the de facto passing of stewardship from the paternal French system to the demands of the U.S.-dominated IMF and World Bank.

As free-trade zones expand and the newly industrializing world rides the roller coaster of global capitalism, the Sahelian countries have been largely excluded. The villages of the Sahelian zone face the future in an extremely brittle macrolevel economic environment, with low growth rates in both agricultural and industrial production and a low per capita GDP. Ninety-one percent of the labor force earns all or part of its income in traditional subsistence agriculture in the rural sector, 7 percent earns its income in government and services, and 2 percent work in mining and manufacturing (USAID 1992). Exports fail to earn the government necessary revenues. A thriving black market bypasses customs controls, and the export of raw materials such as animal skins deprives a crafts sector of value it could add. Nor can Niger compete in world uranium and gold markets, owing to low prices and the high cost of continuing operations given the remoteness of the mines and the country's poor infrastructure.

The relationship between the government of Niger and development organizations has never been overly affectionate. Niger is known to the development community as having one of the most demanding and rigid governmental frameworks to be adhered to in West Africa. There are bureaucratic obstacles in nearly every aspect of program implementation. Officials have been described as being inscrutable and as being as demanding as officials anywhere. Niger is noteworthy for having refused Bob Geldof and his famine-relief Live Aid/Band Aid organization permission in the mid-1980s to use money raised in concerts for development efforts and food aid when they could not satisfy elaborate regulations (Johnson and Johnson 1990).

This burdensome development policy environment has not been alleviated by the threats to national cohesion posed by political crises and social upheaval, particularly among the formally educated students and civil servants in the cities. Coupled with the FCFA devaluation, alienation from compassion-fatigued foreign donors and investors has brought about a condition of permanent crisis for Niger. The bankruptcy of the state has eliminated mobilization campaigns aimed at nation building, including the programs that organized women and youths around public service projects.

IMF-imposed policy changes of structural adjustment were intended to free investment. Yet despite these attempts to unshackle the market, crop and livestock prices have not risen, and declining world commodity prices have meant reduced profits for producers. Poor commodity prices have ravaged the export sectors of many African countries, and closing borders are seen as evidence of indifference toward the vast untrained labor pool. Just as social indicators along the environmental margin rise and fall partially in response to outside stimuli, economic linkages to the outside have profound influence on the remote interior of Niger. Because so many people derive their subsistence partially from migration, Niger's fortunes depend on the economic health of Nigeria and Benin,[1] and stagnation in those countries is at least partially responsible for Niger's plight. Especially in marginal places, where food deficit balances are offset by flows of moving people, development indicators are driven forcibly from without.

Since the 1970s, Niger has not been self-sufficient in food production even in good rainfall years. As Chapter 3 illustrated, increases in agricultural production have been owing to increases in cultivated area, not to technical improvements like higher input use, and the capacity to increase production further without considerable additional investment is extremely limited. To assist the people of Niger in overcoming chronic production shortfalls, donor agencies and foreign governments have provided the country with shipments of food aid. From 1966 to 1973, the U.S. government provided 16,480 metric tons of emergency food aid for seven disasters. From 1973 to 1986, the U.S. Mission in Niamey managed the distribution of 366,520 metric tons of food assistance (USAID 1992), which was channeled by the distribution program Food for Peace into chronic deficit areas such as Ouallam, Tanout, and Diffa.[2]

Since the 1970s, food aid has had a significant impact on rural Nigérien lives. Since the start of the crisis period, observers have noted the questionable benefits of food aid for countries like Niger (De Waal 1988;

Downing 1990). By depressing local grain prices, rewarding corrupt officials, and discouraging investment, food aid erodes the capacity for local
people to adapt, and it is destructive in the long term. Niger's president,
Seyni Kountché, was himself acutely aware of the problems caused by
dependency on food aid and actively sought to reduce the need for it
(USAID 1992). Distribution of food aid in Niger is regarded as a "gift" to
which the better-off and more important have first right. In the region's
political economy, food and development aid emerged in the 1970s as the
new source of accumulation to be captured by politicians and bureau-
crats; food and development aid functioned something like a new cash
crop (Grégoire 1997).

   In response to the droughts of the 1970s, FEWS were implemented in
the Sahelian countries to assist foreign agencies and governments in pre-
dicting in advance where food shortfalls will occur (Hutchinson 1991).
Defined by USAID as a system of data collection to monitor people's access to food in order to provide timely notice when a food crisis threatens, FEWS was conceived and sponsored by the United States through
congressional mandate; it is implemented by USAID (Taylor-Powell
1992). Early warning systems have been coupled with other programs
that seek to bolster households' ability to withstand droughts (Taylor-
Powell 1992). Other than food gifts and food supplements, famine mitigation activities consist of food-for-work[3] or cash-for-work programs, infrastructural development, crop interventions, natural resource
management, and income-generation projects. These activities are aimed
at increasing rural incomes and thereby making famines less likely to
occur, but they are hardly a cure for chronic hunger.

   As an example of development at its unsustainable worst, the showcase Keita Integrated Development Project in Tahoua Department was
funded by the United Nations Food and Agriculture Organization (FAO)
in 1983 to reverse trends of land degradation and "turn back the desert"
in 4,820 square kilometers of Sahelian central Niger. Barren plateaus and
sloping plains of low potential were planted with trees, crops, and grass.
The project constructed roads and schools and implemented social services, conspicuously paying lip service to "local participation." The features of the multimillion-dollar development scheme look attractive and
inspiring in their glossy brochure, but the project met an ignoble end that
only underscores its unsustainability. Owing to region's chronic insecurity, incited by the Tuareg rebellion in the early 1990s, vandalism and
theft (particularly of four-wheel-drive vehicles) have caused much of the
project to be abandoned.

   There remains a lingering sense that such interventions as FEWS, food
aid, and famine mitigation programs are nothing but Band-Aids and that
very few efforts are being directed toward correcting the causes of the

chronic food shortfalls. As it was originally conceived, FEWS has a disturbing surveillance aspect to it, keeping Congress informed about African famines but doing little to reduce long-standing vulnerability in the subject population. If one objective of early warning systems is indeed to keep images of hungry people out of the evening news, then the program can hardly be considered successful.

In the realization that donor largesse was perhaps not addressing the causes of production deficits and was perhaps increasing the problem of dependency in Niger, international efforts have addressed more long-term economic development. France has continued its invisible military and economic supports, which, like most development efforts, are concentrated in the capital city and its environs. France sends its volunteers, and there is a presence of other European countries, including Germany, Belgium, and the Scandinavian countries, which contribute infrastructural aid and work in small-scale projects. The United States has provided support for bilateral health and food security interventions. The Peace Corps has been in Niger for more than thirty-five years and fields volunteers in health, woodless construction, natural resources management, and the African food systems initiative programs. Japan has its own version of the Peace Corps and also donates vehicles and aid. As of 1994, external aid represented 15.2 percent of Niger's GDP (OECD 1994).

As the industrialized world is responding with exhilaration to the economic changes that accompany the opening of new markets and the growth of a middle class, in Niger the absence of such changes is noteworthy. Niger in the late 1990s had an underworld quality to it, with the lack of economic activity here arguably being more significant than the economic activity elsewhere. In towns and villages alike, imported everyday items like coffee, tea, and Bic pens have become prohibitively expensive for many. The exclusion from global markets raises the question, Which is worse, being exploited or being ignored? To approach the question of how the macrolevel context affects settlements and people at the microscale, we must remember how such settlements and people became so marginal in so many ways.

One result of apparent exclusion from a shrinking globe has been the transformation of large parts of inhabited Niger into ecological and economic margins. In the Sahelian zone, where the level of soil maintenance has been low, the practice of fallowing land leaves aside a surplus to accommodate low yields during drought years. What was eliminated in the loss of much fallow and bush land in the Sahelian zone was flexibility in responding to climatic variability. To dependency theorists, marginality in the Sahel is a politically created situation, not a function of the resource inventories of the region. Amin argues that the geographical distribution of factors of production is a result of a strategy of development,

or lack thereof, on the part of the colonial powers. In associating the un-
derdeveloped economy with migration from the interior of West Africa,
Amin contends that there is no need to resort to "pull factors" (Amin
1974a). Inefficient transport, the absence of government help to aid peasants in intensifying agriculture, taxes, and the difficulty in growing and
marketing cash crops all conspire to keep the periphery in a marginal
state. Nowhere are nature or the environment mentioned as factors limiting development.

Yet the margin is both economic and physical. As was illustrated in
Chapter 2, settlement of the Sahelian zone was restricted initially partly
because of political instability. As the region became pacified, settlers
from wetter areas moved in, lured by the prospect of frontier land and re-
assured by the above-average rainfall. At that point, in a sense, the margin was not in existence yet. As long as there was still free bush land to
exploit, the land remained a serviceable outlet for human labor. The
spread of villages and the filling-in of bush and fallow land marked the
human transformation of the region. Increasingly the landscape of the
Sahelian zone became domesticated, or human-managed. What resulted
was better suited for some activities than for others. Specifically, it was
better suited to those that relate *less* to the innate production potential of
the land, or the theoretical capabilities of the soil, and *more* to the land's
relational attributes, particularly in relation to urban population centers
and markets. When the droughts of the early 1970s began to empty the
Sahelian zone, individual peasants utilized the contacts they had developed through trade to ensure their own survival and that of their dependents. These mitigation activities were quite apart from those sponsored
by foreign governments, and they constitute the most historically reliable
coping mechanisms for the alleviation of vulnerability.

The notion of vulnerability as a feature of marginal places must be based
primarily on such socially constructed, locally configured situations. There
is a compelling creative tension between those who wish to privilege
macro- over microlevel forces and those advocating the reverse. At what
scale is vulnerability manifested? Can or should entire regions be considered vulnerable? How does a macrolevel notion of marginality affect local
practices? To answer these questions, a more nuanced understanding of
power as it operates within and between regions must be developed. Here
the example of Maradi is instructive, for it goes beyond convenient polari-
ties and toward a nexus linking the roles of the state, seasonality, and the
institutions of food distribution, including indigenous ones.

To understand the workings of the indigenous food distribution system—for a working system already existed prior to the colonial era—we
must acknowledge just how important trade is in the area, especially
trade in food. The Hausa commercial system is a widespread distribu-

tional network based on a highly elaborate, ritualized, and hierarchical social system of obligation and remuneration. Featured in this social system is an institutional relationship called *bara/uban-gida* (or patron-client). Patron-client relationships between a "master" or "the father of the house" (*uban-gida*) and a "servant" or "dependent" (*bara*) are highly structured and traditional. Servants put themselves at the service of the master, rendering him multiple services without overtly requesting remuneration. In exchange, the master is obligated to offer his dependents gifts. Specifications of the patron-client relationship are determined by perceptions of social importance, of *arzikin mutane* (richness in men); ramifications permeate all facets of the Hausa culture (Grégoire 1992).

This sociopolitical system makes much use of the seasonal flux of people in and out of towns and across ecological zones. Abner Cohen's (1969) study of Sabo (*Sabongari*), a Hausa quarter of the southern Nigerian city of Ibadan in the 1960s, is a treatise on how long-distance trade between savanna and forest belts of West Africa enkindles intensive social interactions among members of different ethnic groupings. The "wagons-drawn-in" conservatism of the *zango* (Hausa enclave) was forged when Hausa colonists placed themselves amid an alien and sometimes hostile environment of the West African coast. Cohen's view is that cultural identity derives not from "tribalism" per se but from vested economic interests that tend to reinforce the power structures of the culture. Cohen's portrait of "the Hausa trader" is a noteworthy:

> His high degree of mobility, skill and shrewdness in business are widely acknowledged and have earned him the reputation of having a special "genius" for trade. On a closer analysis, much of this "genius" turns out to be associated, not with a basic personality trait, but with a highly developed economico-political organization which has been evolved over a long period of time. (Cohen 1969, p. 9)

Such indigenous social structures existed prior to the advent of colonialism. The arrival of European rule did not necessarily destroy existing organizations; rather, they often provided new opportunities for "native elites" to subvert and control. In Maradi, the advent of cash as well as income from smuggling only furthered the acquisition of power among indigenous merchant elites and those who used their own mobility to further their trade interests. According to Grégoire (1992), the maintenance of social networks created more economic power, which in turn made it possible for patrons to redistribute part of their own wealth among their dependents, especially during the growing season, to solidify alliances. Trading networks for groundnuts and other products have followed this structural design.

The city of Maradi is an active site of trade where migrants play a critical role as the labor supply and where seasonal labor and sporadic grain surpluses are exchanged for cash. When shortfalls in precipitation occur, peasants simply do what they have always done: They go to the city, for the city is where the money is. It is also often where the food is. Food imports from Nigeria are not recorded, but they are estimated to be between 80,000 and 160,000 tons per year (USAID 1992).[4] In this process of trading labor and food for cash, migrants gradually become acclimated into social networks that allow them to strike a balance between their family and food-production commitments and their needs for cash. As the value of their time increases and peasants find their efforts to satisfy food needs forcing them out of village routines and into seasonal cash-earning opportunities, the relations with the sending villages are subtly transformed.

Given this urban-based social mechanism, how are present-day responses to drought in Maradi Department different from historical ones? For one, the population is roughly three times bigger than in colonial times, which creates a much larger market for goods and services. Better and more roads mean that goods can be moved more cheaply. A new sense of interrelatedness, brought on partially by radio and television, encourages the spread of ideas. More opportunities makes it easier to earn an income. None of these changes are purely local or global; in fact, they are regional. All together, they amount to an increased urban influence on once-isolated rural villages. This growing influence is marked by an increased level of patron-client relationships, by more trade apprenticeships, and by increased grain speculation and investment.

In the end, it is very difficult to link the global and the local without omitting something or misrepresenting individual and household behaviors. If we understand labor as Marx did, as the medium between people and the land, then there certainly must be more creativity involved in labor's adaptation to drought than a simple choice between starving and fleeing. Nowhere is the relationship between political economy and population mobility more intricate than in the concept of the informal economy, which in Maradi has evolved through the development of both nonagricultural livelihoods and seasonal movements. Urbanization occurs through the movement of archetypal, culturally sanctioned livelihoods out of the village and into new environments. As will become abundantly clear by recounting the history of movement in the region, Maradi has always been relatively peripheral to the world economy, and people have always satisfied their needs through mobility. In the marginal environment of Maradi's Sahelian zone, the caravan (or, in its modern form, the bush taxi) is still the only game in town.

The History of Mobility in the Study Area

Prior to the French conquest, the precolonial economy of the Sahelian zone was based on age-old interactions between grain-growing sedentary farmers and livestock-herding nomadic pastoralists and on long-distance trade between the desert edge and large cities to the south (Lovejoy and Baier 1975). The history of indigenous markets in the Sahelian zone reveals a pattern of incorporation into a regional economy that accommodated the climate through the use of portable production techniques and trade, moving from north to south and back again across rainfall zones. As early accounts confirm, desert-edge products such as grain, skins, cotton, dates, and salt were traded with cities along the West African coast. Trans-Saharan trade in salt, gold, hides, ivory, and dates between Tripoli and Kano is ancient (Davidson 1965). The southward migrations of the Tuareg had begun by A.D. 1,000 and continued over the centuries, thus predating the Atlantic slave trade.

The issue of precolonial economic rationales for seasonal mobility is dealt with at length by historical analyses, which point out the general neglect of the role of commerce in the history of West Africa (Baier 1980; Coquery-Vidrovitch and Lovejoy 1985). Accounts of trade come from European travelers, especially Hugh Clapperton (1829), who accompanied a Hausa caravan passing through Borgu on its return to Kano from Gonja and Ashanti. The caravan consisted of upwards of a thousand men and women and as many beasts of burden. Polly Hill comments on some of the earliest written descriptions of long-distance trade from the Hausa and Bornu empires early in the nineteenth century: "Although the dictionaries define *fatauci* as 'itinerant trading' and although nowadays any man who trades outside his home area is apt to denote himself, however jocularly, as *farke* (long-distance trader), the word is best confined to the type of long-distance trade which was formerly conducted by members of trading-caravans" (Hill 1972, p. 243). Hill contends that in former times most long-distance traders were farmers who were based in the countryside.

By far the best analysis of long distance trade comes from M. B. Duffill and Paul Lovejoy (1985). Their analysis of trade in the central Sudan is based on archival sources and oral informants' testimonies collected in Kano and Katsina. They recount the highly specialized occupation of *fatauci*, which involved a large proportion of the population of the two states of the central Sudan, the Sokoto Caliphate and Borno, in the nineteenth century. Such caravan trade often involved up to 5,000 people and tons of goods for sale, all headloaded (that is, ported atop people's heads) or carried by beasts. Caravan leaders (*madugai*), brokers (*dilalai*), and landlord-merchants operated in a mutually dependent fashion, with the prosperity of the team dependent on the team leader's reputation and skill and the leader in turn dependent on the loyalty and efficiency of his staff, making the operation of a long-distance trade caravan not unlike that of a modern firm. Participation in long-distance trade was regarded as a surefire way to acquire wealth and prestige and to escape from the poverty of *talakawa* (the peasant class). The emphasis on wage labor and entrepreneurship in Hausa folklore confirms the status of trade as a means to gaining riches. Caravan crews shared an identity as workers and acted as a group in times of labor disputes (Duffill and Lovejoy 1985).

When the French arrived in the late 1800s, they somehow imagined they would transform the local economy into a productive machine that would satisfy the needs of the motherland while also turning a profit. The French raised taxes in their *territoire militaire* after 1908 to pay for their new colonial enterprise. Reactions to the head taxes split along ethnic lines within Niger, with a very marked difference between the reactions of the Zarma, on the one hand, and the Hausa, the Fulani, and the Kanuri, on the other (Fuglestad 1983). The young Zarma reacted by migrating seasonally to the Kumasi region of the Gold Coast, following the paths of nineteenth-century warriors. Beginning possibly as early as 1902, these migrations in their early stages involved mainly former slaves but soon spread to other groups. Pointing out the structural similarities between labor migration and long-distance trading, and especially raiding, Fuglestad speculates that among the more "adventurous" Zarma people there occurred a form of cultural transfer that promoted the popularity of migrant labor, to the point where the long journey to the Gold Coast became a part of a modernized initiation ritual (Fuglestad 1983). To Kenneth Swindell (1984), early migration was simply a product of the nineteenth-century economy of the central Sudan, with an overlap rather than a sharp division between precolonial and colonial economies.

In any case, it should be clear that the movements occurring in the colonial era were not introduced during that time but in fact were based on earlier movements that predated the changes imposed by the Europeans. There is of course the hajj to Mecca. But in addition, precolonial movements to the West African coast necessitated the creation of migrant "clubs" in growing coastal cities to accommodate the seasonal flux of migrants, and the *zanguwa* (migrant enclaves) that are the subject of Abner Cohen's landmark study were based on ancient long-distance trading relationships. There were also labor migrations to the North African coasts, to work in the urban economies of Tunis, Algiers, and Tripoli and later in the oil fields of Libya, that were probably inspired by social contacts made during expeditions decades or even centuries earlier. These set the stage for later labor migrations. (See Photo 4.1.)

Photo 4.1   The sign announces the terminal point for the bus line running from Niamey, the capital of Niger, to the large coastal city of Abidjan, Côte d'Ivoire.

In the British colony of Nigeria, dry-season movements among rural farmers had long been seen as an established practice. An observation by W. F. Gowers in 1911 (reported by Swindell 1984) emphasized that long-distance traders were Hausa *farmers.* In French Hausaland, the dramatic increase in head taxes after 1906 and the discontinuation of the use of cowrie shells as currency after 1916 (Fuglestad 1983) only put new stresses on already mobile agricultural producers. In the 1920s, expanding markets in northern Nigeria for groundnuts, cattle, and millet led increasing numbers of Nigérien Hausa people to find work there in the dry season.

The raising of head taxes after 1906 coincided with the end of slavery in the French territory, which released vast numbers of agricultural workers, traders, and laborers. These workers subsequently engaged in dry-season circulation and sought wage employment in areas of both urban expansion and the development of cash crops, including cocoa, coffee, and cotton. According to Swindell, the sharp increases in dry-season migration from northwest Nigeria in the 1930s were mainly the effect of the abolition of domestic slavery and only secondarily the imposition of head taxes and the introduction of a new currency (Swindell 1984).

The French promotion of groundnuts in the 1920s and 1930s allowed the Sahelian population to spread and to grow, and the groundnut boom

financed infrastructure such as roads, marketing boards, and market outlets. This colonial capitalist apparatus fostered the growth of the *el haji*
merchant class,[5] which accumulated capital and state benefits and fed
them into informal urban economies. Ties to countryside through commerce increased throughout the colonial period, and this did not alter
substantively after Niger's independence in 1960.

With European colonization and economic restructuring, historical accounts paint migration not so much as a reaction to changing circumstances but as a transformed and popularized cultural practice. According to Amin, long-distance precolonial trade "has virtually disappeared,
destroyed by the reorientating of West Africa toward Europe" (Amin
1974a, p. 113), but it seems quite apparent that the caravan lives on in
new forms. Clearly, modern migrations did not materialize overnight in
response to newly emerging threats; instead, they followed ancient historical precedent. Thus, historical perspectives do more than trace the ge-
nealogy of the new forms; they also give life to present practices.

When the embryonic states of former French West Africa gained their
independence in the early 1960s, the break was anything but clean. The
French citizens of the colonies retained their business ties in Africa, government advisers stayed in their posts, and a military occupation continued, as did the use of what was essentially French currency. Whether one
sees this as a helping hand or a stranglehold, the good intentions of the
French were inspired to some degree by a feeling of Franco-African *fra-
ternité* that continues to this day. The postcolonial transition had particular consequences for population mobility. Prior to the toppling of the
First Republic of Niger in 1974, President Hamani Diori maintained very
close ties with France and was a particularly strong supporter of relative
autonomy for the country's pastoral population. Diori favored the construction of mobile schools and medical facilities for the Tuareg. This concern with the plight of the nomads met with French approval.

The political economy was disturbed profoundly by the series of
droughts that occurred throughout the Sahel starting in 1968. Hundreds
of thousands of hungry migrants flocked to the cities. The distress sales
of livestock and other assets raised fears of predation by the urban
monied class. Niger's first coup d'état occurred in 1975, after seven years
of below-average rainfall, when Kountché seized power in Niamey.
"Putting the peasant in charge" was Kountché's slogan. In response to
the drought, the state emerged to mediate between the bourgeoisie and
the peasantry and to restore social order. Kountché's Société de
Développement (Development Society), founded in 1975, sought to fos-
ter a new nationalist fervor through a publicly financed social movement
involving traditional groups. This ultimately served the needs of the
state during an era that was marked by a growing dependence on food

and development aid. State coffers were also filled by revenues raised from uranium mining in the north of the country, in an area formerly under the control of nomadic groups. The string of growing towns and villages along the national highway in the far south was evidence of substantial internal migration, which marked the increasing dependence upon outside money for the functioning of the postcolonial political economy.

The sympathy for the pastoralists' need for freedom of movement did not bring about changes in national policy. Development efforts in the 1960s were confined mostly to the southern part of the country, and there was little or no power sharing with the Tuareg and Fulani. Nor were the mobility needs of the sedentary population considered. When Kountché assumed power after the coup in 1974, he enacted a policy of "fixing" the population through restrictions on movement, the issuance of identity cards, the promotion of off-season gardens, and the sedentarization of nomadic populations. Such ideas were associated with Tanja Mamadou, the former *préfet* of Tahoua and minister of the interior under Kountché.[6] Sedentarized Tuareg and Bougajé people were settled in the Department of Tahoua and were known colloquially as *les Tanjawars* (the followers of Tanja). The ill effects of mobility were seen by the government of Niger as the cause of vulnerability in the population, as well as a complication of drought-period stresses.

Kountché ruled with a strong hand and presided over the only blooming of the national economy in the country's history, a brief period of prosperity fueled by the discovery of uranium in the northern part of the country. Schools, clinics, and roads were built, but again they were concentrated in the south, where the sedentary population lived. Antimigration policies remained in effect until 1987, when Kountché died. They were more or less abandoned by Kountché's successor, Ali Saibou, whose tacit policy viewed mobility as a vital part of the Nigérien economy that could not be eradicated. By this point, the migrant clubs in Accra, Abidjan, Lagos, and other coastal cities had filled to accommodate the increased seasonal flow of Nigérien migrants. Penniless and travel-weary migrants could go to the mosque in practically any coastal city to pray, and there they could meet other Sahelians who could house them and possibly find them jobs. Instead of relying on the postal system to send money, migrants used friends, family, and an informal *cantine populaire* (informal savings organization) system to funnel remittance money back to villages in Niger. In the early 1990s there were an estimated 100,000 permanent residents from Niger in Abidjan (Painter 1992).

This explains why remote villages in the Sahelian interior of Niger often have electrical generators, refrigerators, and televisions. Throughout West Africa the presence in rural homes of expensive goods like radios, fashionable furniture, and wall clocks is often a sign of migration income (Adepoju 1974). Seasonal migration assumed its importance in Tahoua in the 1980s and 1990s, and the villages of Tahoua in central Niger are supported in a considerable measure by remitted money. Up to 80 percent of men aged sixteen to fifty participate in seasonal migrations, as do a smaller percentage of divorced women who have left forced marriages in their villages. According to Hada Goga, who runs a CARE project in Tahoua to promote condom use among migrants, approximately 80,000 migrants from Tahoua Department migrate to the West African coast annually, with most going to Abidjan and residing in *zanguwa* (strangers' enclaves) in the city. Remittances range from nothing to more than 200,000 FCFA[7] each year, most of which is sent back to the villages of Tahoua Department and spent on millet, animals, and clothing. Migration remittances are also invested in housing and agriculture, including field labor. Perhaps more than 50 percent of the local economy is supported by migration remittances. When plotted by income, the migration prevalence among Tahoua residents seems to be bimodal, with wealthy and impoverished residents migrating more often, and middle-income residents more likely to stay home in the dry season. Tahoua has been called *"la ville d'exode par excellence* [city of seasonal migration without parallel]."

Tahoua's dependence on remittances is sustainable in the short term in that individual migrants can successfully strike a balance between home and opportunity, but the future might be less promising. When I visited Tahoua in October 1995, 170,000 migrants were being threatened with expulsion from Algeria. One of Mu'ammar Gadhafi's periodic threats to expel all the Nigérien migrants from the oil fields of Libya may yet come to pass.

The most tangible threat to international migrants from Niger is from AIDS and other sexually transmitted diseases. There has been a rapid increase in cases since 1987 to a total of 1,729 at the end of 1994. Ninety percent of the men with reported cases of AIDS are migrants, and 90 percent of seropositive women are reported to be married to those who migrate to Abidjan (*New York Times*, February 8, 1996). During 1994, 62.7 percent of those with AIDS were found to reside in Tahoua Department, where the Pilot Project for AIDS and Migration in Niger is located. Yet despite the threat of AIDS, migration to and from the coast continues to grow to unprecedented levels. Yet in the final analysis, mobility to and from Tahoua is plainly a case where the benefits for the migrants outweigh the risks.

The growth of the multiparty democracy movement and resultant explosion of political party constituencies in the late 1980s and early 1990s had specific consequences for Niger's migrants. There was a realization that the Nigérien migrant population represented a sizable force that could be mobilized for political ends, and this realization was compounded by the government of Niger's tendency to lock any potentially profitable social development into a stultifying bureaucracy. The creation of migrants' associations in the early 1990s was therefore a direct result of the convergence of state political and economic interests, which underscores the importance of population mobility for the region. Amadou Karijo, then the civilian *préfet* of Maradi, explained that the associations were intended to encourage aid to the villages. Efforts were made to register migrants and to enlist their help in sending money to their homes. But the result of this mobilization was that the migrants' associations quickly became recruitment tools of the political parties and a way for them to gain converts in rural areas. Once the associations' political aims became clear, the government pulled its official sponsorship and financial support. Now, if they exist at all, the migrants' associations are purely party recruiting organizations. Rarely in my interviews with migrants did anyone admit to belonging to the migrants' associations, and many migrants thought the associations had been outlawed by the government.

The contemporary history of Niger's politics and policies toward mobility have their counterparts in many sub-Saharan African nation-states. Clearly the precursors to present-day population movements predate colonial policies and colonial boundaries. Because there are relatively few barriers inhibiting movement, people flow relatively unhindered by government, bypassing military or customs controls. Given the level of population movement, the very concept of residence denotes something more fluid than in the West. Frequently the "places of mobility" are based on opportunities, not on static locations alone.

The implications of these two points—weak barriers to movement and a widespread emphasis on opportunity—are manifold. Much population redistribution in Africa can occur in the coming years without adverse impacts, but only if people are allowed to relocate based on their calculated needs. The transition from rural to urban can be peaceful, productive, and without incident as long as two conditions are met: The social fabric must stay intact, and migrants should be encouraged to continue growing food. These are prima facie sound conditions to test, and to balance, against the negative consequences of so many moving people. In fact, the founding charter of the Economic Community of West African States (ECOWAS), which was signed and ratified in the late 1970s, recognizes the need to maintain, if not encourage, intraregional migration as a way of rationalizing and optimizing resource use at the regional level (Ezenwe 1983, p. 192). Article 27,1 of the charter states that all obstacles to freedom of movement should be abolished. Why then has this not informed development policy?

In the development programs operating in Niger, seasonal migration is seen as an early indicator of household stress; estimations of seasonal migration form an element of FEWS vulnerability assessments. As a survival strategy for use in below-average rainfall years, mobility is seen as having a specific geography in Niger, which corresponds to agro-ecological zones. The northern Sahelo-Saharan zone is dominated by pastoralists and is based on herd mobility, raiding, and age-old self-governance mechanisms. The southern Sudano-Guinean zone has good agricultural potential and seems to show a willingness to intensify agriculture. In this zone, mobility is used to secure agricultural materials, to funnel investments into villages, and to gain better access to markets. Niger's Sahelian middle belt, from fourteen to sixteen degrees north, is the area likely to be agriculturally unsustainable in the long term because rainfall is too variable and the soil is too fragile to permit continuous cultivation. Field sizes are large, and yields are unpredictable. It is here in this transition zone where seasonal migration might tip the balance between vulnerability and preparedness for drought and famine. This situation characterizes Tahoua, with its heavy dependence on money from the outside and the unclear impacts of this dependence on long-term sustainability.

And yet migration policies in the Sahel are covered with a political veil because development policies are driven by a bias toward permanence, which leads development actors and national governments to act in the interest of preserving state stability at all costs. Development actors point accusatory fingers at mobility, even though it is based on ancient practices. As an indicator of poor rural conditions, mobility is associated with the failure of development efforts. And since the developmentalist solution to poor conditions is more development, the solution devised to improve village conditions is a reduction in mobility. Such a mind-set calls for people to stop moving, settle down, and invest more in their villages. This "sticky places" fallacy—that with the proper development interventions, the conditions can be created to allow people to stay put—reveals the contradiction of mobility and the conflicting roles mobility plays in this region. As will be seen in the next section, mobility acts as the vehicle that allows the rural sector to remain viable even as it is decried as an indicator of vulnerability.

Mobility as Economic Enabler and the Death of the "Dead Season"

The antimobility mind-set flies in the face of observed behavior in Niger. Clearly mobility is being made to work in rural areas; it shortens distances and make circulation of goods easier. As Jane Hopkins, an agricultural economist from the International Food Policy Research Institute (IFPRI), has argued, the income from seasonal migration is financing investments in agricultural production and sustainable change (Hopkins and Reardon 1993). Hopkins regards migration as a form of resource circularity, in that the labor lost to migration is replaced by flows of other goods, especially goods to enhance long-term sustainability of agriculture.

Hopkins's and Reardon's study of rural income sources echoes the conclusions reached by other researchers in Niger, that in spite of conventional images of the Nigérien producer as being self-sufficient and subsistence-oriented, rural households are in fact engaged in an active monetized economy (Taylor-Powell 1992). Emerging new opportunities arise from markets and trade, and urban market demand is already transforming production systems near urban centers. Urban populations are expected to increase between fivefold and eightfold by 2025. These realities raise potential implications for seasonal migration and might enable marginal settlements to continue to existence by becoming more active players in rural-urban networks.

The economic data the IFPRI/Institute National de Recherches Agronomique du Niger (INRAN) team collected are the most precise ever for Niger. The team found that crop sales accounted for about 10 to 25 percent of income, with an average of 15 percent. Data on incomes collected in 1989 found annual household income in these rural areas to be about 37,000 FCFA (US$148)[8] per adult equivalent, on average.[9] Household incomes included revenues from households' cropping their own lands, agricultural wage labor, animal husbandry, commerce, transport, construction, food preparation, gathering, small-scale manufacturing, migration income, and other cash transfers. In cereal equivalents, the team found that about 14,000 FCFA (US$56) per adult equivalent per year met minimum needs.

The IFPRI/INRAN team found a clear inverse correlation between agro-climatic potential and the prevalence of both short (three to five months) and long (six to twelve months) seasonal migration. Migration was highest in the driest study area, northern Boboye, with 28 percent of the adult male population gone for all or part of the dry season. Hopkins and Reardon also found that income from migration formed up to 20 percent of total household income in the northern study areas and 2–6 percent in the southern ones. Most migration income is brought back to the village at the start of field preparation and planting, as opposed to being sent back by postal money orders (which accounted for less than 10 percent of money sent back). Intersectoral income links such as the financing of migration through crop sales or the use of migration earnings for on-farm investment are very common.

When broken down by income tercile, the data for the northern (Sahelo-Sudanian) zone reveals that migration income played an important role in overall household income. The level of migration was seven and a half times higher among upper-tercile households than among low-tercile households, and two and a half times higher than middle-tercile households. The share of total income more than doubles, from 10 percent to 22 percent, between the lower and upper tercile. This demonstrates the high barriers to entry for migration. The need for bus or bush taxi fares, seed money for profitable microenterprises like textile resales, materials for crafts, and support for family members who are left behind (Taylor-Powell 1992) puts long-distance migration out of reach for many villagers.

In all regions except the Sahelian northern Boboye study area, female income was greater than or equal to the value of noncereal, nonlivestock purchases, which included agricultural purchases such as meat, milk, fruit, and vegetables, and nonagricultural purchases such as health and educational expenses. Since female income is frequently used as seed money for male migration, a web of gender relations is revealed that may have implications for how Nigérien households operate.

What the researchers found to be most significant was that overall rural household incomes did not vary by agro-climatic potential. In areas of lower production, income derived from cropping was augmented by nonagricultural income, including migration remittances. This finding implies that in the villages studied, migration remittances serve the perceived needs of the village for the maintenance of a subsistence minimum, and this highlights the village regulatory function performed by absent village members. Information about the well-being of family and friends is transmitted through a social web connecting absent members with the village. During times of crisis, news about the village reaches the absent ones in the cities. Money or food is sent home, and the village recovers. In a sense, this social web connecting the village and the city enables a kind of remote control in which *just enough* resources are provided to keep a village functioning, but not enough resources are supplied to allow the village to thrive. This form of surveillance is made possible by large numbers of migrants, who are no longer present in the village on a full-time basis but who nevertheless maintain a watch. Through the mechanisms of circular migration and social networks, food security in the rural sector is therefore maintained at a minimum level of sufficiency. Where crop production potential is lower owing to climate and soils, household economies diversify to bolster household income, using circular mobility in many cases. Set against the agricultural calendar, the development of nonagricultural livelihoods constitutes an adaptation to changing circumstances.

In investigating economic opportunity in the village, Hopkins and Reardon point out that "dead season" is an inappropriate term and that projects and programs that rely on farmers being idle or having low opportunity costs in the dry season will come up against the fact that farm families are very active in the dry season and earn possibly half of their yearly income then. This diversification varies by zone, with the driest receiving the largest percentage of nonagricultural income.

If we conceptualize labor in the context of Sahelian conditions of economic production, then for the study region, the investment in a particular place actually means investment in the people of a place. For as long as the area has been occupied by cultivators, there has been a season when agricultural labor needs are low. These needs also vary from year to year. Depending on the previous year's rainfall, there has been wage labor in the hot season, clearing and burning brush. Across the Sahelian zone from July to October, men are generally engaged in work in the fields.

For the residents of Guidan Wari, a main economic resource is the agricultural fields, and the primary occupation of men is farming (Jones 1995). Men can make money from their fields by selling off the surplus, if there is any, as well as by buying up surplus in a time of plenty and selling it for profit when times are harder. The major crops for Guidan Wari are sorghum, millet, and cowpeas. There has not been a good millet crop for many years, but in 1995 the sorghum and cowpeas did well, so well in fact that many families sold and shipped cowpeas to the city. Millet and sorghum are consumed in most cases by the family and their animals. From these raw food sources, women make and sell food for profit. In addition, merchants and skilled laborers in the village make money by selling their goods and skills. After the grain is dried and harvested, laborers usually spend some weeks in activities such as making bricks, weaving fences, and repairing roofs.

Women can ostensibly buy what they want with the money they earn. In principle, the husband is responsible for all the basic needs of the family, such as cooking pots and extra clothing, for example, in order that the wife's (or wives') money can go toward nonessentials. However, many times the family needs the money that the wife makes in order to meet even basic needs. Likewise, children start participating in household economics as soon as they can work—at about seven years old—to support the household. Children will do light work in the fields, feed and watch over animals, sell food, run errands, care for younger children, help draw water, and do whatever else they are asked, most often without direct compensation.[10]

Though historical evidence is less than complete, the increased level of monetized rural activities probably only began in this century, and possibly in the 1980s and 1990s. In Barth's account, outside of the agricultural season, the villagers of the region have "very little employment at all, with the exception of weaving a little cotton" (Barth 1857, p. 431). This transformation of rural activities, particularly through short-distance and short-duration movements, makes sense if the land cannot be relied upon and if the ties to the land itself were never very firm. Mobility's vital role in accessing off-farm incomes makes its role for the migrants a benign one.

We must not lose sight of where occupations are performed in relation to natural resources. Of the occupations already discussed, some—such as herding—require land. Some—such as pottery and masonry—require clay. And others—like crafts, weaving, and smithing—require specialized supplies. But many occupations require few resources other than those provided by the people themselves. With the exception of herding, none requires a sizable parcel of land. This in effect makes mobility practicable: There is no purely economic reason to stay settled. Because ties to the village group remain strong over distance, most who decide to go away briefly do not completely alienate themselves. But as we will see, the children of the migrants find their affinities becoming more urbanized, and this is how the social context for urbanization evolves. This can help explain why policies to staunch the flow of migrants are bound to fail, because nowhere is the need to have better local infrastructure to make occupations viable addressed.

The historical process that is apparently occurring in the villages of the Sahelian zone, owing to changes beyond the control of rural villagers, is the transfer of these "portable" occupations out of the villages. In other words, what was formerly done in rural settings to fulfill daily necessities has became monetized and has thereby fallen under the sway of the urban social universe. In the past, when cash needs were lower, villagers could use the off-season to rebuild their houses, make pots, and weave mats. Whether circular migration in the Sahelian zone is seen as depriving villages of labor that could be used to make them more viable economically or whether the migrants themselves are seen as proof of the lack of demand for this labor depends on how unified a village is thought to be. At a certain juncture, the needs for cash overran the ability of the local economy to provide, and people started to look to the urban markets for their rural occupations.

The relationship between the indigenous institution of the market and the infrastructural changes put in place by the French calls into question our ability to distinguish between cash and subsistence sectors. Despite the forced conversion of the economy at the hands of the French, it would be misleading to suppose that prior to colonial rule there was no taxation or cash economy. It might be more accurate to regard subsistence and cash economies as two ends of a continuum, not a dualism. In fact, there are likely to be cash exchanges *and* noncash exchanges occurring simultaneously within a single household. These are used to bind individuals and households together into a web of mutual obligations, which has less to do with the actual cash value of the transaction than with the bonds that are cemented. Hopkins found that in the Sahelian zone, crop sales did not correlate with agro-climatic potential; most occurred within the *arrondissement*, and half were to other farmers within the same village. In most study areas, the most common stated reason provided for crop sales was to buy food (Hopkins and Reardon 1993). Household members explained that they are nearly as willing to transfer crops as to sell them, and this underscores the continuing importance of social relationships in the village.

In discussing the role of money in Hausa culture, Polly Hill notes that Hausa people "do not share the pejorative European attitude toward cash as filthy lucre. Cash is a positive good, and there is no reason why a personal relationship should create any inhibitions about giving or receiving it" (Hill 1972, p. 29). The intense circulation of goods and wealth all along the commercial networks has as a corollary in the penetration of a system of monetary exchange that reaches to the heart of the rural world. In this context, the image of a peasant society living in a state of pure subsistence is totally erroneous. Commerce is an ancient activity in the Hausa culture. Studies of villages in Maradi in the 1970s and 1980s found salaried farming and commerce in prepared foods was routinely practiced, even in isolated bush villages (Grégoire and Raynaut 1980). Market movements of cereals, beans, livestock, and groundnuts occurred mainly from the north to the south, whereas imported goods moved from the south to the north. Of course, the importance of motorized transport (especially trucks) in the circulation of goods has had a considerable effect on the commercial spread of consumer goods into the rural areas. To gauge accurately the consequences of the urban hub of Maradi on the rural areas, we must return to the city and connect the idea of *detachable* components of the village economy with the *seasonal* urban demand of informal sector activities.

Owing to Maradi's late start and the parallel late development of the Sahelian zone of Maradi Department, it is very difficult to isolate the rural sector from the effects of the urban market. Most of the colonial infrastructure was built in the 1920s when the area north of the fourteenth parallel was being settled. The embryonic road network including road links east, west, and south began in 1921, and telephone service between Maradi and Zinder and the installation of European commercial houses[11] developed at the same time as the rural areas were settled. Thus, the rural settlements were tied to Maradi's success or failure from the beginning.

Urbanization and urban-rural relations in Maradi reveal the pervasive influence of urban capital. When groundnut cultivation spread across the region in the colonial era, the colonial power built a distributional network atop a system of hierarchical commercial relationships that already existed in Hausa culture. The urban hierarchy linked venture capital to the rural production system and included Hausa *el haji*s and Lebanese and French entrepreneurs. Relations between merchants and farmers were conducted through intermediaries, who distributed gifts and advances to the influential persons in each village. These gifts and advances were frequently distributed during the growing season (June through September) when farmers needed money. Interestingly, these intermediaries were named *madugai*, after the word used for the caravan leaders in the old cross-Saharan trade system.

With the 1968–1974 drought came a halt in groundnut production, and the city filled with famine victims looking for food or work. One attempt by the government to reverse the flow was the "return to home village" campaign of 1974, when the military used trucks to return drought refugees to the bush, with each load accompanied by a load of food. Yet the old commercial network remained, wounded by the collapse of groundnut production but still actively involved in alcohol, cigarette, and gasoline smuggling. Today the *masu tebur* (petty roadside sellers) who sell their wares virtually everywhere in Maradi Department—even in remote rural areas—probably got their goods on credit from a merchant in the city of Maradi.

It is therefore quite difficult to see the vaguely sentimental "tribal" affinities between buyers and sellers, or among friends, as anything but hard economic rationality at work. Etched permanently into the Hausa culture through the time- and space-specific accumulated actions occurring over space, the long-distance trading caravan was the archetype for a host of later forms of mobility-based economic relationships. It is in this way that we see the culture of movement as specific to the region where it is practiced. As both the sum and the consequence of actions occurring continuously in the same place over time, culture is not itself tied to static places. The participants in the caravan, even if they are sedentary farmers during planting and harvesting times, are relentlessly, daringly mobile; they are masters of the portable place. Their place affinity is an ephemeral feeling of belonging that can be toted about and thrown up like a tent. Permanence as Westerners know it, encased in marble palaces and inscribed on granite tombstones, rarely exists in this environment. Monuments to permanence are replaced by more durable though locationally indeterminate affiliations and obligations—family and lineage ties, alliances, enmities, and trust—that trail back into and out of remembered history. Residence in the landscape of settlement is fleeting and temporary, an improvised existence on a convertible stage. But as the stability of "permanent temporariness" of the arrangement makes itself apparent, urbanization is altering the relationship between the city and the village, with consequences for the latter.

The City Wins: Urban-Rural Contests
over Land, Food, and Labor

While the villages in the Sahelian zone slowly adapt to the endogenous changes in their midst, the city gains influence and progressively consumes the region. A remarkable feature of Maradi is the high percentage of urban residents who are still self-sufficient in food they have grown themselves. Grégoire (1992) notes that an estimated 70 to 80 percent of Maradi's urban households produce their own food, both to eat and to sell, and further reports that everyone does a little farming in the rainy season.

The reciprocal relationship between agriculture and mobility is demonstrated in the growth of periurban farming. Villagers in nearby villages rent their land to urban "gentlemen farmers" for growing food crops. Adamou, a schoolteacher who has lived in Maradi for ten years, grows his food on a large rented field about thirty kilometers west of the city, near Goulbin Maradi. Villagers are apparently happy to rent the land because the soil gets fertilized, Adamou surmises. Adamou rides his motorbike out to supervise the planting and cultivating in his rented field. In all, he budgets about 100,000 FCFA for farming every year, with 10,000 FCFA going to rent the land and 45,000 FCFA going for two sacks of fertilizer, which is bought in Katsina with bribes at the border. Adamou pays laborers 500 FCFA per day per person to plant and weed the field. Last year, the land yielded eight 100-kilogram sacks of millet and fifteen 100-kilogram sacks of groundnuts, which he stored in the back room of his house in Maradi. In the period before the rainy season when the prices are at their highest, Adamou sells his surplus and pockets a hefty profit.

The need for cash among an increasingly mobile peasant labor force, especially at crunch times, marks the growing prevalence of wage labor within the region. This was illustrated for me by Ado, a man in his early twenties who works as a launderer in Maradi. He embodies the role played by circular migrants within Maradi's economic system. According to Ado, "The *masu digga* [temporarily displaced] are hungry and leave their villages after the rains finish, and there is, as usual, a poor harvest so they are destitute." Ado made derisive comments about young men who are unable to plan ahead and then end up eating all their millet seed.

> In the city they do odd jobs and at least support themselves, but after six or eight months they have no savings to show for their work. After the rains start, they go back to their villages to farm, but they have no seeds or fertilizer—nothing but the land. So they do wage labor in their village or in other

villages, working for the wealthier village members, doing their hoeing, planting, and weeding. They work for 500–600 FCFA per day, sometimes as much as 750–800 a day, sometimes as little as 250 per day.

It should be noted that the wage rate depends on the available labor supply, which is in turn governed by growing conditions the previous year. In these instances, the relationship between landlord and migrant worker is governed by tradition. Normally the landlord is obligated to feed the laborers who work in his fields. This saves the laborers money, which they can use to buy their own millet seed. This type of wage labor is called *yawon kwadago* (wet-season labor) or *barema* (begging). Again, as with other forms of mobility, the traditions are old but the popularity is more recent. In Guidan Wari, the old chief confirmed that there always were people working as day laborers on others' fields, and this year there is not enough money so there will be lots of labor. And if there is a bad crop this year, there will be a large pool of available labor, and this will mean that more land in the area—probably the last remaining bush land—will be converted into fields.

In the bigger villages in the south of the region, where the lower drought risk and higher land values make investments more common, trade in foodstuffs is controlled by groups of wealthy investors. According to Ado, these men know the prices of cowpeas, millet, chickens, sheep, and beef at all the nearby markets. If they hear that a good is selling for less at a distant market, they will go up there or send someone to buy it out. Ado spoke very admiringly of those who had not only the means but also the foresight to make such investments: "You make 125 francs on something. Now the 'uncivilized' would eat it all, or eat 100 FCFA and put 25 FCFA into savings or investment. But the smart investors work the other way around, eating zero or 25 and investing 100, and that way their money can grow." He told stories about people he knew in Maradi, the sons of *el hajis*, who had no business sense and who lost money on cars or flashy consumer goods.

Clearly there are those who benefit from the seasonal fluctuation in food prices, and for those who play the game well, gambling on the arrival of the rains and the start of the agricultural season can be very lucrative. In these cases, circular migrants themselves form a fluid medium between economic interests in the village and opportunities in the city. The upshot is that food is increasingly becoming a traded commodity, and rural peasants are becoming, effectively, the minions of a commercial class of investors.

This metropolitanization of mobile labor will only continue. Claude Raynaut (1988) sees the future of Maradi hinging on government decisions, with levels of future migration depending on the agricultural policies adopted by government planners. The choice lies between either set-ting aside security stocks of grain to limit the commercialization of pro-duction or continuing the creation of an entrepreneurial class of highly productive farmers. The latter, Raynaut argues, will accelerate the rural exodus. On the contrary, one could argue that this view runs counter to local realities, particularly regarding contests over land. Can peasants be "uprooted" in a region where mobility is endemic and land is commu-nally controlled? More likely is the continued spread of urban conditions that make increasing use of labor mobility while retaining ties to the land. This process of economic transformation is actually responsive to regional change and integral to the development of the region. It does not—and should not—depend on government policies or graft and cor-ruption to work. But such a change carries a price: In the short term there is balance in this system, but over time the city gradually wins out be-cause people stay there. In the end, the city wins in every way.

While Raynaut's and others' fears about the exploitation of desperate peasants by the bourgeoisie may be justified, treating all migrants as des-perate is simply inaccurate. Who is more vulnerable: those who leave, or those who are left behind? Increasingly the villages on the outskirts of Maradi are being supported by migrants' cash and gifts, partly to ensure that the migrants can continue growing food there. For more distant vil-lages in drier zones, the data on migration remittances seem to suggest a monitoring function at work. Research on the role of migration income in the rural economy shows that the amount of migration income in a vil-lage overall corresponds to the amount of income needed to bring the vil-lage up to a minimum level of sufficiency in foodstuffs, with per capita village income likely to be uniform across agro-ecological zones. This would explain why migration levels are more pronounced in areas of marginal production potential.

This surveillance dynamic raises an intriguing issue. If villages *are* in fact being supported from the outside as the data suggest, then the level of circular migration over time should correspond to the magnitude of opportunity in the city as well as to production conditions in the rural areas. Contrary to Amin's thesis that modern migrations effectively op-erate without a recourse to "pull factors," in fact migrants respond to both the absence of potential in the village and the presence of urban op-portunity.

Mobility, Environment, and the
Political Ecology of Hazards

Misunderstandings about the historical role of commerce in Hausaland are on display in a research paradigm devised in the 1980s. Political ecology is characterized as blending the concerns of ecology and a broadly defined political economy with an emphasis on the constantly shifting dialectic between society and land-based resources and also within classes and groups within society itself (Blaikie and Brookfield 1987). For the purposes here, the emphasis on state intervention in the rural economies and the differential responses by decisionmakers to changing social relations of production and exchange are noteworthy. The generally Marxian orientation of political ecology maintains that people's vulnerability is caused by powerful social agents. Famine, created and exacerbated by the international capitalist economy, is regarded as a consequence of food commodity production for the world market. Population impacts such as hunger migration are conceptualized mainly as responses to outside stimuli and are not seen to be governed by their own internal logic.

In application to the analysis of production conditions, this approach has some weaknesses that bear on the present case. Social power at levels of geographical scale below the region is inadequately conceived. Even with the black box of the household opened, what can be glimpsed inside does not explain everything. There are limited historical perspectives on how individual production units change over time. Uncertainties exist as well in the interface between human activities and their environmental effects occurring at the community level of analysis. These shortcomings are compounded by an interaction-averse perspective on local processes, which treats outside encounters as perversions of a so-called pure subsistence system.

This view of politics and hazards has taken residence in international development and cultural studies. The funneling of capital, raw materials, abundant foods, and manual labor from backward zones is seen as fueling the rapid development of growth poles and as effectively relegating supply areas to increasing stagnation and underdevelopment. With respect to Sahelian West Africa, Amin (1974b) and Raynaut (1988) describe "marginals" created by colonial capitalism who, in the process of withstanding conflict and the reallocation of space, were pushed into less-desirable places. Setting up housekeeping in these marginal spaces, the in-migrants found they could only precariously maintain their culture. Trying to maintain their culture along the margin, the inhabitants became increasingly vulnerable to seasonal shocks.

This process seems straightforward, and yet it has often been bound in ideological and polemical debate. Nowhere is this debate more pronounced than over natural hazards, where few cool heads have prevailed. It is a given that there are places and regions where people are more vulnerable than in others. These risky environments are near the faults, on the slopes, by the tracks, or in otherwise less-advantageous places. The old school of thought on natural hazards (for example, Burton, Kates, and White 1978) asked why people moved to and lived in hazardous places, and its answer was constrained decisionmaking and the lack of knowledge about risks. Whereas the old school of thought on hazards blamed the "bounded rationality" of the peasant for moving into risky environments, new hazards theory, which is based on the political economy approach, essentially blames "the system" for making people vulnerable. Because social relationships are organized on the basis of unequal power, they always distort the effects of natural disasters. Richard Franke and Barbara Chasin (1980) portrayed the great Sahelian drought, and the outpouring of concern around the world, as being built upon a sketchy understanding of dryland ecology and on developmentalist greed. Others more recently have reapplied this political-economic argument and argued for the relative unimportance of drought as an underlying *cause* of famine; rather, they argue, it acts simply as a trigger (Blaikie et al. 1994). Through the web of contingent relations, including cash crop production, taxation, forced labor, household pressures, land clearance, and lack of grain reserves, a long causal chain is created that ultimately lays the blame for famine at the feet of colonial capitalism.

The paradigm of political ecology has one of its most weighty proponents in Michael Watts, who is influential within academic geography and Third World development studies. He introduced Hausaland to Anglo-American audiences with *Silent Violence* (Watts 1983a), a massive work that historically links climate and society in northern Nigeria. Watts asks why famines occur and how their effects have changed through time. His answer amounts to a single chain of exploitation paradigm, a chain in which famines were and are organically linked to the rupture of the balance between peasant subsistence and consumption, precipitated by the development and intensification of commodity production. More concretely, Watts maintains that subsistence crises in Nigerian Hausaland are grounded in the unfolding of capitalism in Nigeria.

The period of European colonial rule in Nigeria and the Nigerian petroleum boom of the 1970s are worthy targets for Watts's commentary. Yet the leaders of the crisis-prone political monster who subverted national will and diverted oil money into corrupt and foolish schemes are not presented as responsible for the tragic conditions that have developed there. In Watts's view, it was capitalism itself that caused the twentieth-century famines that plagued Hausa peasant society. Watts prefers to sketch out the broader and more dramatic story of the transformation of peasant economies, effectively downplaying precapitalist crises. Prior to the colonial era, the normal risks of agricultural production could be shouldered through the strengths of social relations of production, a kind of moral economy (Scott 1976). During this period, famines were merely precapitalist crises reflecting the obvious technical limitations of the productive system. By expanding commodity production and individuating peasant society, capitalist development in northern Nigeria ruptured the cycle of peasant production. During postcolonial crises, peasants were subject to the horrors and moodiness of the market without the benefits of transformed forces of production. After the capitalist transformation, there was apparently little a village could do to prepare for a famine.

Presenting such an overarching set of theoretical linkages imbues power to the narrative, but never is evidence presented that colonial and postcolonial famines themselves differed substantively from those of pre-colonial times. Watts states as fact that in the nineteenth century, the Sokoto Caliphate (in what would later become northern Nigeria) exhibited a remarkable resiliency to climatic stress, but this is not documented through testimonies, physical records, or historical accounts from the time. There is in fact ample evidence of a long history of famines and epidemics prior to the capitalist era in the West African Sahel, a history that parallels that of the droughts. The parade of famines that can be drawn from historical sources (Gado 1993) is relentless prior to colonial rule. Famines occurred in 1657–1660, 1669, 1670, 1695, 1711 (lasting seven years), 1738, 1741–1744, 1758, and 1762–1766. Then there were twenty years of respite at the end of the eighteenth century. During the time of the Sokoto Caliphate, five Sahelian famines were recorded, in 1833–1834, 1864–1865, 1867–1870,[12] 1880–1881, 1884, 1889, and 1890–1892. Plainly, there were severe droughts and subsequent famines during the period in question. If the impacts of such famines were somehow lessened given the village moral economy and social relations of production—and this would certainly be unusual given the high mortality rates and the slow natural increase of population during precolonial times—Watts never provides any evidence to substantiate this.

As Watts himself documents, trade and labor in Hausaland were and still are fairly inseparable, and taxes,[13] forced labor, slavery, and cash crops all existed prior to colonial rule. How then do precolonial and colonial or postcolonial famines differ? One could argue that famine is actually caused by a combination of exogenously *and* endogenously derived stress, which is brought on by the limitations of the productive system and exacerbated by destructive and misinformed colonial policy. Instead of being a fitting subject for a grand narrative, the reality of famine in the region is actually rather mundane. Pre- and postcolonial famines may differ not because of colonial rule itself but because the world in which famines occur has changed so dramatically. One such change is the growth of infrastructure and social networks,[14] which make severe famines less likely to occur but may also raise the level of vulnerability overall in a village, particularly for those unable to move to mitigate the effects of drought.

What is perhaps most misleading is the implication, which Watts does little to dispel, that capitalism in Hausaland is a colonial import. In fact, Islamic capitalism is woven into Hausa society and has been part of Hausa society as far back as the sixteenth century. A lingua franca of West Africa, the Hausa language (and culture) itself spread and developed through itinerant travel and trade. The commercial component of civilization is independent of the later French and British influences, which used but did not invent the social system that they found when they encountered the civilization at the end of the nineteenth century. The implications for population mobility are plain: Mobility has been woven into the fabric of Hausa society for centuries, and all contemporary relations and exchanges reflect this fact.

Notes

1. The interdependence runs both ways. A study by the government of Niger (République du Niger 1990) of the 1988 census found that of the 15,330 declared immigrants in Maradi Department, 13,085 (or 85.3 percent) were from Nigeria.

2. In addition to the U.S. government, major donors traditionally involved in the emergency food assistance program have been the World Food Program, the European Development Fund, France, the Federal Republic of Germany, and Japan. In drought times, Belgium, Italy, Pakistan, Saudi Arabia, Libya, Algeria, and Nigeria have also sent emergency shipments of food, as have private voluntary organizations (PVOs) such as the Red Cross/Red Crescent, CARE, and SOS Sahel (USAID 1992).

3. According to USAID (1992), food-for-work programs enable famine victims to support themselves when traditional coping mechanisms are critically stretched. Food-for-work and cash-for-work programs are targeted at food-insecure areas, where hungry local people perform day labor in natural resource management and similar activities in return for food.

4. Millet and rice tend to be imported from Nigeria, whereas cowpeas and onions are exported there.

5. The title *el haji* refers to someone who has made the pilgrimage to Mecca. The pilgrimage (or hajj) is one of five required duties of Muslims, and the respect afforded to those who are wealthy and devout enough to make the hajj is considerable.

6. Tanja Mamadou is now head of the Mouvement National pour la Société de Développement (MNSD) political party and was one of three candidates jailed during the June 1996 presidential contest.

7. In 1995, 200,000 FCFA were worth approximately US$400, based on an exchange rate of 500 FCFA for one U.S. dollar. This is the rate that will be used throughout the study.

8. It should be noted that these incomes were measured prior to the FCFA devaluation in 1994.

9. Adult equivalents were calculated to account for variations in size and composition of households, by weighting the age and sex composition by coefficients

standardizing the person to a proportion of an adult man in terms of caloric requirements (Hopkins and Reardon 1993, pp. 21–23). The ratio of inactive (age less than fifteen and greater than sixty) to active household members ranged in their means from 1.3 to 1.5.

10. Within the family, work is divided among the wife or wives, the husband, male children, and female children. Men do certain kinds of work in the fields and support the upkeep of the house. Women do more harvesting of crops, preparing crops for cooking or selling, cooking, drawing water, caring for children, and tending to animals. Female children help their mothers in their work and tend to younger children. Male children have few responsibilities, but they may do some work in the fields or animal care as they grow older (Jones 1995).

11. The Compagnie du Niger Français (CNF), the Société Commerciale de l'Ouest Afrique (SCOA), and the Compagnie Française de l'Afrique de l'Ouest (CFAO) all handled the export of groundnuts and sold consumer goods.

12. The 1867–1870 famine is referred to as *ci-korya,* "eating calabashes."

13. Evidence of precolonial taxation in the form of a 10 percent grain levy (*za-kkat*) is mentioned by Watts himself (p. 68), along with the fact that exchange was intrinsic, as were patron-client relationships.

14. "Le facteur essential qui différencie les crises alimentaires d'hier et celles d'aujourd'hui est l'existence d'un réseau routier et des moyens de transport permettant de joindre facilement les régions déficitaires [The essential factor that differentiates yesterday's food crises and those of today is the existence of a road network and modes of transportation that permit easy access to deficit regions]" (Gado 1993, p. 47).

5

A Population in Motion

Once the rains end in late September or early October, the millet and sorghum crops are left to dry in their fields for a few weeks, and then they are gathered. Whole unhulled sheaves of millet are bundled and placed in granaries and storerooms, which are closed for the year. Surplus grain that is destined to be sold is threshed by women, and other produce including cowpeas are also prepared for market. The months of October and November are the peak of market activity for the year. Once the harvest is finished, villagers repair the damage done by the rain to houses and compound walls, and they attend to other neglected household tasks. After all these tasks are completed, many villagers' thoughts turn to leaving the village.

The journey to the city begins with a long walk or a whole-day wait for a bush taxi to fill. In Guidan Wari, the preferred vehicles are ancient gray crank-started Land Rovers. On a market day they sit on a path in the village in the late afternoon, after five o'clock prayers, waiting to fill. The driver waits with friends in the shade while the apprentices collect the 500 FCFA fare (US$1), and gradually the vehicle fills. Passengers grapple with the truck sides to find seats atop the cargo, which at this time of the year is perhaps twenty 100-kilogram sacks of millet, as well as chickens, guinea hens, and loads of miscellany. Women with babies or young children sit toward the center of the cargo to avoid falling off the moving vehicle.

The bush taxi ride from Guidan Wari to Maradi is long and slow owing to the poor quality of the roads. For most of the route the road is rutted sand tracks running between millet fields. Because these vehicles are weighed down by tons of millet and beans, with the human and animal passengers on top of that load, the truck usually experiences several flat tires. Once several overheated inner tubes are replaced and a few challenging sand ruts are negotiated, the truck is well on its way to the city, arriving by way of back roads to avoid the police stop outside of town

Photo 5.1   The loaded Land Rover en route from Guidan Wari to Maradi demonstrates how overstressed the transportation infrastructure is between rural villages and lower-order market towns. Roads and vehicles have seen dramatic increases in use in recent decades.

and the inevitable bribes that are required to pass through. The truck enters the darkened city at two o'clock in the morning. This journey of almost 100 kilometers has taken ten hours (Photo 5.1).

In any given truck full of passengers from Guidan Wari, depending on the season, there are passengers traveling to Maradi-*ville* to sell grain or other goods at the market; these passengers will stay over the weekend to take advantage of the Monday market day and then return home. Others have various other motivations and destinations in mind. There are men traveling to Maradi to work for a few weeks or a few months. There are people returning to Maradi, their customary residence for most of the year, after spending the rainy season in their village to work in their fields. There are also some passing through on their way to Nigeria.

As we have seen, the movements made by the passengers on the bush taxi represent in fact a *circulatory system* based on familiarity and social contacts, choreographed to the pattern of rainfall. As a result of the heightened movement between city and countryside, the Maradi region is increasingly becoming a unified economic organism. As cities have grown in size and importance, more and more daily and weekly movements between nearby villages and urban hubs have occurred. The effect of this parallel development has been a convergence of interests and an embrace between town and country. The periodic movements away from the village represent a step in the transformation of the rural world and have spawned a *culture routière* (road culture) across northern Hausaland.

This chapter and Chapter 6 are devoted to the Maradi migrants. This chapter will present the results of the survey-interviews conducted with

TABLE 5.1   Guidan Wari Mobility Types

Mobility Type	Local (Hausa) Name	Practitioners	Duration	Destinations
Temporary displacement	*digga*	men	2 weeks to 6 months	border cities, Nigeria
Seasonal displacement	*cin rani*	men and women	3 to 6 months	border cities, Nigeria
Short-term wage labor	*barema*	men	1 week or more	nearby villages
Market mobility	*kasuwa*	men	1 day to a week	local markets
Educational mobility	*karatu*	young men and boys	6 to 9 months	regional cities

seasonal migrants in the city of Maradi. I will begin by examining the identity of the migrants and their different social characteristics. The nature of contacts and visits between migrants and the rural villages, and the process of becoming linked into the urban economy and social infrastructure, will be investigated after that. Finally, I will return to questions about the continuity of the Maradi migrants' social worlds, as expressed through their stated level of satisfaction with the city and their future outlook as urban residents.

Local Mobilities and the Urban Economy

In Guidan Wari, I discussed with villagers the population mobilities prevalent in the area. Asked to evaluate existing typologies, the informants produced their own in addition, which is presented in Table 5.1. According to the informants (who incidentally were all male), *masu digga*[1] are in the category of temporarily displaced. *Masu digga* generally have little available cash, and according to the informants this catches them in a seasonal bind. Migrants have no money and little food, yet they must save their seed stock for planting. Their conventional motivation for traveling is to reduce consumption of the household food supply, but paradoxically, having them go away usually does not save food because many do not eat at home with their spouse(s) and children anyway, preferring the company of brothers or male friends in the village. They may, however, send money back home from their work in the city, which can be used to buy grain and other food. The duration of these temporary displacements is from a week or two to several months. The informants maintain that these temporary displacements are often distress moves, driven by household food shortages.

The *masu digga* can be contrasted with *masu cin rani*, or the seasonally displaced, who often consist of nuclear families—as opposed to complex households—who leave the village after the harvest and head toward urban centers in Niger or in foreign countries within Africa to look for work. Migrants from Guidan Wari usually take up residence in Maradi or in Jibiya, across the border in Nigeria. The duration of such movements is usually around six months, after which time the migrants return to their villages. In their off-season home, it is possible for everyone in the family to find work. The man can participate in the informal economy, the woman can pound grain or do other food-preparation activities, and the children can sell things or do odd jobs. Together they can make adequate money to invest, perhaps, in livestock. But even if this does not happen and they do not bring windfall profits back to the village, they still have saved all the food they would have consumed in the village. The informants were of the opinion that the *masu cin rani* were the migrant success stories of the village. One farmer who took his family to Jibiya did not make enough to invest, but every year he and his wife returned fatter and healthier than they would have been if they had stayed at home in the village. According to the informants, the seasonal displacements are usually *not* distress moves.

Such mobility activities are most prevalent among the *talakawa* (peasant class). Asked if people with money, either the *sarautu* (royalty) or the Islamic bourgeoisie, practice either *cin rani* or *digga*, the informants replied that people with means invest in the market rather than migrating. This produces in effect a third local mobility type, called *yawon kasuwa* (walk of the market), which suits the dispersed settlement pattern of Maradi Department.[2] Guidan Wari's Friday market draws people on foot and on donkey from twenty or more kilometers away, especially since Friday is also the day for prayers at the mosque.

Wage labor is common at planting and harvest time, when those without money work in other people's fields. The informants termed this form of activity *barema*, from the word *bara* (begging). The explanation provided was that the region is marked by women's participation in agriculture, and poorer households rely heavily on women's labor in the fields, which creates a potentially difficult situation for women with small children. Instead of using women's labor, wealthier households hire men instead to alleviate the labor bottlenecks. The informants identify this tradition as older than anyone in the village can remember, and as old as the settlements themselves.

The fifth mobility type prevalent locally is *karatu* (Koranic training), whereby young men leave the villages for Islamic study with *marabouts* in Mayahi, Maradi, and elsewhere. Koranic teacher-scholars have regional importance and attract casual and long-term followers (*almajirani*). The modern educational system in Niger has modified this Islamic system only slightly, and the young are indoctrinated into mobile ways at an early age, sometimes as young as twelve or thirteen in the case of *sixième* (first-year middle-school students), who sometimes travel 100 kilometers to board at and attend a *collège d'enseignement général*. By leaving their village, they are exposed to the greater world and experience an often irreversible change in values.

The mobility category of transhumance represents a traditional economic activity of the pastoral population in Niger, which traditionally raises livestock. This category is not represented in the typology just outlined for Guidan Wari because even though most residents are originally Bougajé pastoralists, Guidan Wari is a sedentary village. Livestock herders often bypass the village, preferring to water their animals where there is more space. In addition, "pure" pastoralism is becoming more rare, with many individuals abandoning their herds and now working in the city for wages.

The existence of exceptional categories led me to examine the local mobility typology. Both the seasonally and the temporarily displaced correspond directly with a typology developed by Niger's planning office. Not included in the Niger typology were *barema* (wage labor), *kasuwa* (market), and *karatu* (Koranic study) types, which are apparently not considered serious migration because the distances covered are short. Also, the Niger planning office category of permanent displacement (*kaura*) did not appear in the Guidan Wari typology. An individual who is initially a seasonally displaced migrant may succeed in a new location and decide not to return to the village the following year. It is quite likely that because the migrants are not presumed to break with the village definitively, permanent displacement is not widely recognized as a separate category from the villagers' perspective. Villagers always prefer to think that long-absent members will soon return.

Female migrants in the Guidan Wari typology were also not represented in their own category. Marriage mobility was not considered by the Guidan Wari villagers as a separate mobility type, owing perhaps to the localized nature of marital bonds, with many individuals marrying from within the area. Fulani, Zarma, or Hausa women may migrate, either in the company of husbands or without them. Divorcées, who are known in Hausa as *karuwai* (free women), are well-known urban residents who often work as prostitutes or who prepare foods for sale or engage in petty trading.

In the urban economy in Maradi, seasonal population fluxes provide the labor for microenterprises and the service economy, and the twice-

Photo 5.2   A shantytown inside the city of Maradi on the grounds of an old hippodrome shows how much the urban settlements resemble their rural counterparts. Informal housing uses the same recycled materials and uses the same forms of governance as well.

weekly market plays a crucial role in organizing contacts. Migrants are most often found in quarters of millet-stalk houses on the outskirts of town, on open-access and rent-free land. One such shantytown was built on the grounds of the old colonial-era hippodrome (Photo 5.2), but it has been left alone by the *préfet* since the early 1990s. Such communities in the city look very much like villages, and they are governed in a fashion similar to recognized *quartiers*, with a kind of provisional authority by the longest-term residents. The army and municipal government keep their hands off these neighborhoods for the most part, although some residents expressed fear of eventual eviction. The laws on squatting are ambiguous in Niger, and there would be political consequences to removing migrant camps.

Owing to its size and its proximity to the border, Maradi has a ramshackle vibrancy that is lacking in some other Nigérien cities. Noteworthy sectors include the old town, which dates to 1945 and features mud houses in close contact with each other, hugging the ridge above the seasonal river, the Goulbin Maradi. This section of town (see Figure 5.1) contains its long-term residents, with few recent migrants.[3] Just as security was a limitation to colonizing the Sahelian zone, likewise the city from its inception had limits set to its growth. After the 1945 flood forced some residents to move out to the bush or to Nigeria, the town was relocated to the ridge top, and it began to grow, only becoming a trade center after the French established themselves.

The administrative town, with the *préfecture*, the post office, hospital, and *gendarmerie* (army barracks) lies at the hub of the spokelike street layout. The market and *tasha* (taxi-park) lie along the main road, which con-

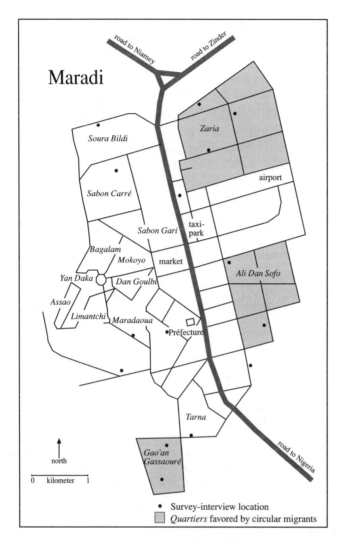

FIGURE 5.1   *Maradi*-Ville, *with Migrant* Quartiers

nects Niger to the rest of the world, with the north road veering left for Niamey to the west and right for Zinder to the east. The south road leads to Nigeria. With goods everywhere and taxis filling up constantly, the main road to Nigeria is a continuous string of social contact points, which allows people to survey the vehicles heading to Nigerian cities like Katsina and Kano. There is the same active roadside in other Nigérien border cities such as Dosso, Konni, Maradi and Magaria, permitting a free flow of information between those in cars and those on the roadside. Those merchants along the road are in excellent positions to reach potential customers and also to act as information brokers. The new neighborhoods (*sabon garuwa*), which were built to house the thousands who moved into Maradi in the late 1970s, lie to the north and east of town. These areas have wider streets and feel more spread out. Migrants also occupy shantytowns on the outskirts of the city.

The city economy is based in the central marketplace and in the streets, where hundreds of microenterprises are located. Street corner congregation and commerce is the rule in all parts of town, including the administrative area. The market operates on Mondays and Fridays and draws thousands of day visitors from the region, but the most significant dimension for marketing activities is seasonal.

Maradi Migrants as a Sample of a Population

Questions and doubts arose as Kanta and I were interviewing our "hidden population" of migrants as to whether this was one population or in fact several different populations.[4] In order to winnow out those who had definitively left their villages, we settled on a working definition of migrant as someone who still travels back to the village on a *regular* basis, to visit or to practice agriculture. In a sense, everyone in Maradi is a migrant, because nearly everyone is from somewhere else, and they or their families have located in Maradi within the last few generations. Intentions are sometimes different from actual behavior; sometimes migrants can be gone from their villages of origin for years. By using the criterion of having traveled back to the village in the past year, it was relatively easy to focus on those migrants who still spend significant amount of time there. Civil servants, teachers, and other more permanent Maradi residents were largely not included, but guards, with their sporadic employment were, because their residence in the city follows a seasonal cycle that allows them to have their feet planted in both urban and rural places throughout the year.

The group of Maradi migrants we interviewed was 89 percent male and 11 percent female. A masculine migration flow is typical of West Africa (Zachariah et al. 1980; Oucho and Gould 1993), even if one considers female mobility for the purpose of marriage to be a form of short-distance migration.[5] Female participation is more common in rural-rural migration. Rural-urban circular migration is far more occupation-driven. A point made by one of the urban informants related to gender roles in the city is instructive. Compared to life in the village, the urban lifestyles tend to reduce the traditional differences between men and women. In the city, women participate in the market and administrative sectors, they drive cars, they ride bicycles, and they buy and sell at the market. For them, the urban realm is a stage for surpassing the constricting social norms and roles in the village, though at present, relatively few vocational possibilities in the city are open to women.

In their age distribution, the Maradi migrants are somewhat typical of migrating populations around the world (Clark 1986), with a heavy concentration in the twenty- to thirty-five-year-old age range and a peak in the twenty-five-to-twenty-nine range. Migrants I interviewed had a median age of thirty-five, with peak age categories between twenty-five and thirty-nine making up over 50 percent of migrants surveyed. In another study of seasonal and temporary migration conducted in Niger, the median age of seasonal migrants was found to be around twenty-nine years (Dankoussou et al. 1975). Why are Maradi's migrants slightly older than the global median? I speculate that this has to do with Maradi's function as a gateway city and with a life-course explanation for mobility destinations. Because traveling seasonally to northern Nigerian cities is strenuous and is usually reserved for younger men, many of the migrants I interviewed had long since decided to stay in Maradi. The age distribution displays a slight rise around age fifty, reflecting these career seasonal migrants who have settled more or less permanently in Maradi. I recognized the age group that was eluding me when by chance I met a young farmer from the outskirts of Maradi who was leaving the next day for Kano, Nigeria, to seek work. Many of the residents who simply pass through Maradi en route to work opportunities in Nigeria were not easily found, and since these more enterprising migrants tend to be younger, it explains why I was able to find so many older migrants so much more easily.

Demographically, Maradi migrants mirror Nigérien society at large. Like the general population, the migrants tend to be married or had been married prior to migration. Among the men in the group, 86.6 percent were married, 13.4 percent were single, and none was divorced. The figure for males compares with the general Nigérien adult population, in which 72.8 percent of men are currently married (UN 1979).[6] Among the fourteen female migrants I interviewed, three were married and eleven were divorced. This contrasts sharply with the overall female adult population, in which 82.7 percent of women are currently married (UN 1979). Of the male migrants whom Michael Mortimore interviewed in the late 1970s, 65 percent were married with children. Again, the fact that the number of single men migrating is relatively small here may be related to destination. Maradi contrasts with the larger cities on the coast, which attract younger and more ambitious migrants.

On their most recent trip to Maradi, a majority of migrants walked or rode alone. In all, 57.9 percent came alone, 39.1 percent were accompanied by their families, and less than 1 percent were accompanied by a friend or came with a parent. Of the respondents who arrived accompanied by their family, especially small children, some implied that they were in distress.[7] Of those who came alone, 37.7 percent were rejoined by their families in Maradi later, and of those more than half were rejoined within a year. Some respondents reported that they were rejoined by their wives and children several years later, in some cases ten or more years later. This compares with the findings of other studies. Only 16 percent of the male migrants whom Mortimore interviewed brought their wives and children along on the migration. Eighteen percent had no wife or children, and a further 17 percent had no children (Mortimore 1989). A sizable number of Maradi migrants (15.8 percent) reported that they met their spouses (usually first wives, sometimes their second wives) in Maradi. Most of the women interviewed did not meet their husbands in Maradi.

In FEWS reports, the presence of wives and children in seasonal migration streams is interpreted as indicating distress. This interpretation was supported at times and contradicted at others. Many of the youngest migrants made their first trip to the city while traveling with their parents. Older migrants tended to take their families along on migrations. Divorcing women sometimes took their children when leaving their village, although sometimes they were forced to leave their children behind.

Though the number of dependents varies by age, the migrants are rather typical in regard to family structure in Niger. Kanta speculated to me at one point that men with larger families were more likely to migrate, since feeding all their children twelve months out of the year in the village was impossible. Among the migrants, the average household size is 6.5; this mirrors the average size of 6.4 persons per household in the overall Nigérien population, in which over 30 percent of households have 6 or more dependent children (UN 1995). The average number of children in a sample of households in Maradi Department, according to the Niger Demographic and Health Survey, is 4.3 (DHS 1992).

Migrant interviews about this dynamic reinforce the validity of the explanation that migration was partially influenced by the pressure to feed families. Bara Abou,[8] age thirty, assessed his decision to migrate as positive, "despite the problems of feeding a big family without enough money." When asked why they had migrated, a very common response given by those migrants in Maradi with large families was, "I can feed and support my family." Malam Harouna Issa, the *marabout* whom we met earlier, made his first migration on foot from Chadakori, fifteen kilometers from Maradi, with his wife and their eleven children in 1975 because of food shortages in his village. The children begged in the city for money and food. He states as his reason for residential satisfaction that his children can eat. It should not be seen as coincidental that the *marabouts*, who often are among the most vociferous opponents of family planning in Niger, tend to have very large families.

It is through the social process of habituation that children learn the urban life; and over the generations, that is how Hausa culture has come to span so easily both city and country. Through exposure to school and a new environment, obligations to the old village dissipate after a generation or two. This is a significant sociocultural implication of mobility that we will revisit later.

In their level of formal education, seasonal migrants are also representative of the Nigérien society at large, being only slightly more schooled than the general population. According to the Niger Demographic and Health Survey, 89 percent of male respondents have no formal education (DHS 1992). Migrants' overwhelming illiteracy and lack of formal education leave the impression that they are being omitted from a civil system that has made only the faintest gestures toward including them.

Asked why they migrated *the first time*, 83.4 percent gave food, money, or food *and* money as the reason. All three responses stem from a decision made after the harvest to maximize food stocks and to attain self-sufficiency in the city. Physical or environmental forcing factors themselves seem to be weak as justifications for migration; only two respondents gave drought as a reason to migrate, and two suggested that shortage of pasture played a role. The National Migration Survey in Burkina (Cordell, Gregory, and Piché 1996) found that over 80 percent of men (and 13 percent of women) interviewed offered money as the motive. Only female Maradi migrants cited divorce or marital problems as their reason to migrate. Three percent left for religious reasons, to pursue Koranic study.

For the age distribution of the first migration for the respondents, again there is a typical age profile for African migrants. The rite-of-passage trigger explains a large peak around ages fifteen to nineteen. Over three-quarters of the respondents left the village for the first time before the age of twenty-five. For the age-specificity of the migration activity, the case of Garba Garba is typical. A *Bori* priest from Dakoro, Garba left his village at age eighteen when he walked to Madaoua, and then he took a bus directly to Abidjan in Côte d'Ivoire. For eleven years, he sold traditional medicines in ten different cities all over West Africa. He reported that after he made the hajj to Mecca, something changed inside him, and he realized again that he was Nigérien. He returned to Maradi and settled there, founding a new shantytown on the outskirts of the southern part of Maradi. He has lived there for seventeen years and has raised a large family. He seemed settled, but when I interviewed him in Novem-

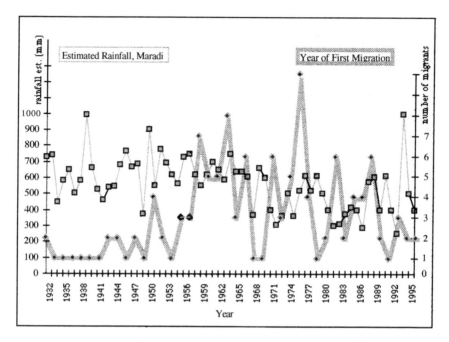

FIGURE 5.2   Rainfall and First Year of Migration

ber 1995, he thought it might be time to move on. He said he thought the city was getting too crowded. He also expressed fear that the authorities would soon come and clear the area. Gao, the shantytown *quartier* where he lives, is known around Maradi as a den of thieves, smugglers, and gamblers. Garba fears that the bad reputation will provoke a police crackdown in the neighborhood.

Correlating the year of the first migration with precipitation totals reveals drought to be an influential trigger for migration. Figure 5.2 charts the variations in rainfall in Maradi Department with the years of first migration for the Maradi migrants. The peaks in migration correspond to the 1968–1974 and 1982–1984 droughts. The parallels between rainfall deficits and decisions to migrate are easy to imagine. Low cereal stocks during or after the drought prompt people to move. In all, almost half of the Maradi migrants left during or around (one year or less before or after) a drought. This compares with the GRID (1990) study, which found that 28.1 percent of their sample of 242 left in 1975–1980, supporting the drought-trigger explanation. As mentioned in Chapter 4, drought years correspond to societal crisis in Niger, including political or economic crises. With 91 percent of the economy based on agriculture, this is not surprising.

Questions relating to migration motivations were asked to determine the extent to which the most recent decision to migrate was made individually or by the head of household. In response, 47.4 percent said that migrating was their "own idea," 21.8 percent cited family, and 21.1 percent cited friends. Due to the diffused nature of decisionmaking, it is difficult to separate bona fide personal decisions from those motivated by other family members. This produces what is likely to be a slight overestimation of those who said they made the decision to migrate for their own reasons. Some migrants replied that they were motivated by an old migrant, that they went to Maradi for Koranic study, or that they accompanied their spouse. Other respondents stated they left their village with their animals, or they left for school. One respondent said that the army chased him out of his village. Work motivations often figure in the decision to migrate. Three responded that a *patron* (sponsor) provided the idea, and another stated that an opportunity to practice his trade in the city was what motivated him.

The question of whether the migrant's family accepted his or her decision to migrate was intended to draw out possible household conflicts over the subject of migration, as for instance whether a spouse, children, or parents object to a migrant's leaving. In the responses, this kind of conflict was absent or not expressed. In all, 82.7 percent of the migrants replied that their families did accept their decision to migrate, and 10.5 percent replied that their families did not accept their decision. In this case, as with the others related to decisionmaking, it is difficult to ascertain the exact role of the family, meaning in some cases the respondent's father, in influencing the migrant's decision to migrate.

In part, questions about transportation and travel were asked to ascertain the costs of moving to the city. Many migrants are from what are, in effect, the outskirts of Maradi, traveling only a short distance. Almost half (45.9 percent) traveled by bus, automobile, or truck; 39.8 percent came on foot; and 10.5 percent arrived via a combination of foot and car. Many coming by foot are from villages within a day or two's travel of Maradi, and most had to walk a short distance anyway to the nearest bush taxi. The costs of the travel likewise tended to be relatively low.[9] Walking is popular because the travel distances are not very great and also because most migrants are so cash-poor that they cannot afford the bush taxi fare. Some who travel on bush taxis beg for free fare, pleading poverty, or they work on the vehicle loading and unloading cargo. Working on the lumbering, heavily decorated trucks is a common way for young men to see Nigeria for the first time.

Questions about trip cost were asked to determine the level of investment made by the migrant in the move to the city. Most moves are short and inexpensive, though even 500 FCFA might represent a sizable sum to a hard-pressed villager. The question about who or what financed the trip was asked to determine the level of planning for this migration, to see to what extent it was a conscious household strategy. For those who paid cash for their transportation, a sizable portion (27.9 percent) stated that they sold grain or other foods in order to finance their move. This is often facilitated by millet or livestock sales. Still others relied on family, friends, charity, or indentured work to pay for the journey to the city.

The issue of planning or premeditation also figured in the next two questions. The question about which itinerary the migrant followed was asked to detect the practice of migrating from smaller to larger cities in a steplike fashion. The questions about knowing someone in the city prior to migrating and about lodging and feeding were designed to isolate the role of social networks for the migrants.

Most of the Maradi migrants (60.9 percent) in fact came directly to Maradi or after a few days' walk with a *sha ruwa* (water-drinking) stop or two in between. About 39 percent traveled in several steps, reaching other destinations before Maradi. Most (66.2 percent) of those who had visited before also knew someone in Maradi prior to migrating. This reveals the social network function that works to the benefit of first-time migrants. Very few remain anonymous in the city for very long, either after visiting or after a more permanent move. Resident family members are very important in providing a smooth transition from village to city, and respondents cited big brothers especially as being crucial.[10] Many respondents answered that they relied on friends or on people from their village or region to help them after they arrived. Even so, a quarter of the total migrants relied only on themselves for food and lodging after their arrival in the city. This presupposes a modicum of cash to buy food, at least for the first few days.

Those who relied on themselves for food and lodging when they arrived in Maradi sometimes had already accumulated some savings from migration to other destinations. Laouali Maman, age thirty-six, first left his village in 1973 and walked from Dakoro to Maradi. Then he traveled to Kano, where he worked as an apprentice for seven years. He had money in his pocket for food when he returned to Maradi. He now makes a living selling black market gas from Nigeria, using the connections he developed while living there. He did not rent or buy a house but instead built a straw house on free land in Ali Dan Sofo *quartier.*

The question about how long a migrant has been in Maradi is deceptive, for it points out the ambiguity of residence for so many circular migrants. The process of becoming acquainted with the social networks that provide opportunities, frequently in the informal sector, follows from both knowledge and experience. Unlike Todaro's characterization of urban migrants waiting around for modern-sector work, the nonformally educated migrants in Maradi pound the streets upon arrival looking for temporary employment. As they get older and gain more experience, the migrants become more adept at finding opportunities that allow them to earn cash *and* to maintain their contact with the village. It is through their determination to create a livelihood for themselves that migrants get absorbed into the urban economy. The tenure of most migrants is rather long, with a substantial portion of respondents residing in the city for five years or more.

The role played by repeated visits to Maradi in making it a desirable destination for seasonal migrants is evidenced by the fact that 72.2 percent of current migrants stated that they had been in Maradi before. The motivations for repeated migration behavior highlight the differences between the first-time migrants and veterans. First-time migrants left their villages following tradition, because their fathers left, or they were forced out by drought or they went with their families, either as a child or as an adult. Over the years a process of learning and self-selection occurs, and the migrant develops what demographer Douglas Massey (1987) calls "migration capital," which is the knowledge about destinations, routes, way stations, and opportunities that allow migrants to improve their well-being while out of the home village. Attaining this is normally a process occurring over many years, in which accumulated place-specific knowledge plays an important role. Bearing in mind Niger's economic dependence on its neighbor to the south, it is not surprising that more migrants travel from Maradi to Kano in northern Nigeria than to Niger's own capital, Niamey. Kano is physically closer than Niamey, and the border is less of a barrier than the distance. It is surprising how few migrants from Maradi Department choose coastal cities such as Abidjan. Hausa-speaking northern Nigeria is perhaps more welcoming to them.

Maradi's gateway phenomenon in seasonal migration has a notable effect on the geography of settlement in southern Niger. Since most of the seasonal population movements are from north to south (and back again), towns and villages appear to form in chains, with specific paired linkages running perpendicular to rainfall distribution. The directionality of these movements reflects the routes of both livestock corridors and the old long-distance caravan routes. Variation in migration destinations creates certain "staging cities," with the middle city of the chain being the gateway city that "holds" a lower-order service sector employed by a more mobile and affluent population. These better-off people travel to the southernmost city, where the economy is most vibrant and where there are more work opportunities. Examples of these linked cities include Dakoro and Maradi in Niger and Katsina in Nigeria; Tahoua and Birni n'Konni in Niger and Sokoto in Nigeria; Tanout and Zinder in Niger and Kano in Nigeria; and Filingué and Gaya in Niger and Malanville in Benin.

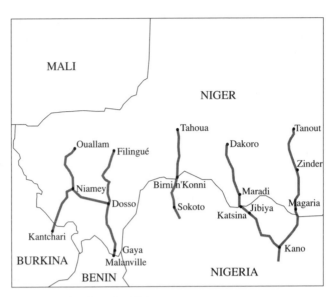

FIGURE 5.3   Gateway Cities

This phenomenon is illustrated in Figure 5.3. It is likely that a similar urban network based on the dynamic of movement perpendicular to rainfall gradients is prevalent in other parts of the Sahel.

Individual migration histories were collected retrospectively. Many of the moves can be described as itinerant. For example, migrants travel to Sokoto and look around for work there or beg on the streets for food. Either they find work or food and stay, or they leave empty-handed for another destination. So the length of stay is an indication of how successful the migrants are. Figure 5.4 is a map of the destinations of migrants interviewed in Maradi, based on migration histories collected from the migrants and on the number of times individual cities were mentioned.

The migration history of Rabi Moussa is typical: Over the course of twenty years, he has visited the Nigerian cities of Lagos, Ibadan, Shagamu, Oyo, Abomosho, Ogo, Wari, Ona Cha, Oma Apia, Inabou, Atoukou, Makourbi, Lafia Beriberi, Boko Birni, Katsina, Okari, Teletaraba, Ibé, and Okeni, working during these stays variously as a soldier in the Nigerian army, a cigarette vendor, a shoeshiner, a tea vendor, and a merchant. He typically stayed from a few days to three to four months, but sometimes, depending on the work activity in the city, a stay would last an entire year. In most years he returned to his village of Indotokalgon Kwassaou in Guidan Roumdji *arrondissement.* In years when he did not return, his older brother was there to cultivate the family fields.

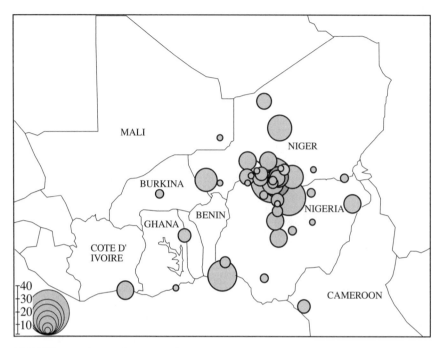

FIGURE 5.4   Seasonal Migrants' Destinations

I asked about Maradi as a destination to ascertain which attributes of the city make living and working there rewarding and also to gauge the level of planning. This was an open-ended question, and we took care to record the responses as they were given and to classify them afterwards. By far the most common responses to the question about Maradi's "pull" relate either to the proximity of the city to the migrant's village of origin or to the size of the city. Migrants from within Maradi Department can balance their dry-season work in the city with their social and farming obligations in the village. In the words of one migrant, "I can't go any farther than Maradi," without losing the advantages both of the city and the village. Another migrant, age twenty-seven, chose Maradi to test whether it worked to be away from home before going further. Many other respondents stated that they "know Maradi" and that their familiarity with the city helped them navigate through its networks. Other respondents cited social, personal, or family ties as the reason for choosing it. Table 5.2 lists the reasons given for choosing Maradi.

Abdourrahaman Idi, who has been in and out of Maradi for six years, said he knows where to find work here now. A female respondent, who

TABLE 5.2   Reasons Given for Choosing Maradi as a Migration Destination

Reason for Choosing Maradi	n	%
Proximity and size	48	36.1%
Social/personal/family ties	22	16.5%
Food availability	15	11.3%
Other reasons	51	36.1%
	133	100.0%

works in Maradi as a prostitute, chose to move from a nearby village in Nigeria to Maradi because she wanted her seven-year-old son to recover his health. In other cases, either the migrant came to visit family in the city and decided to stay, or an opportunity presented itself to work with members of the village of origin living in the city. Others stated that food availability, work availability, the intervention of God, or the undesirability of other migration destinations (namely Nigeria) persuaded them to move to Maradi. Of all the responses, it is surprising how infrequently individuals rationalized their decision to move based on village conditions. Few stated that their decision was forced. Far more common were responses justified on individual or personal grounds, which in Table 5.2 have been lumped into the category of "other reasons." In all, I got a strong sense that most migrants exercised a considerable amount of choice in their migration destinations. Further research in other cities would be useful to see if the same answers occur.

Questions about housing type and tenure were asked to probe the changes in the migrants' material worlds when they moved into the city. The "village in the city" describes well the migrants' settlement in Maradi. Almost 65 percent live in straw houses, and about half as many live in mud-brick houses. Housing types change very little in the transition from rural to urban. A majority (84.2 percent) of seasonal migrants are squatters, living for free on open-access city land, which was frequently converted from other uses. Some occupy space in the shadows of more permanent housing or in the compounds of unoccupied dwellings. Those who pay rent often share the expense with other working family members. Of all the migrants surveyed, only three identified themselves as owners of their own housing. Worldwide, migrants are three or four times more likely to be renters than property owners (Clark 1986).

Social networks are critical in finding work for the migrants. Occupations recorded in the 1995 interviews are listed by category in Figure 5.5. It should be noted here that many of these are actually traditional *rural* occupations. Many are in fact portable and use social networks to gain access to the informal economy.

Community leaders	Farming	Services
community organizer		apprentice driver
marabout	**Guards**	baker
		barber
Crafts/semi-skilled	**Market/sales**	brickmaker
butcher	cigarette sales	car rental
apprentice mason	clothes seller	cook
carpenter	condiment seller	driver
griot (praise-singer)	frozen food sales	laborer
hatmaker	furniture sales	launderer
Hausa bed maker	gas sales	water carrier
traditional medicine	peanut seller	porter
mason	table sales	shoe repair
sandal maker		shoeshiner
snake charmer	**Post-harvest**	singer
sword maker	fence maker	water carrier
tailor	rope maker	

Not employed	Women's occupations
beggar	food preparation
unemployed	milk seller
retired	millet pounder
	prostitute

FIGURE 5.5   *Migrants' Occupations*

Earnings in the city were derived from the respondents' own figures, often estimated over the week or month. As the experience of many migrants shows, it is possible to subsist in the city on the equivalent of less than a dollar per day (500 FCFA); 58 percent of the respondents estimate that they earn 500 FCFA or less per day. On the average, circular migrants in Maradi earn 763 FCFA (approximately US$1.50) per day. Assuming this wage remains steady throughout the year,[11] this equals US$547 per year, which is about twice as high as the national per capita GDP of US$270.

To find work in the city, respondents used both family and acquaintance connections. These social networks are clearly responsible for the often quick transition to working life in the city. Table 5.3 lists the meth-

TABLE 5.3   Method of Finding Work

Method of Finding Work	n	%
Friends in city	48	36.1%
Asked around	11	8.3%
No work	8	6.0%
General family	7	5.3%
Koranic study	6	4.5%
Networks	5	3.8%
Big brother	5	3.8%
Father	4	3.0%
Relied on self	3	2.3%
Apprenticed	2	1.5%
Other responses	34	25.4%
	133	100.0%

ods used for finding work. (The variation in earnings and the income distribution in relation to occupation type, age of the migrant, and level of experience will be covered in more depth in Chapter 6.)

It is accepted that migrants can make more in the city than they could in their villages. Many maintain that this is what motivated them. But the size of the migrants' earnings compared with, or in terms of, their buying power is not entirely clear. This question was asked to see how many of the migrants earn enough money to send some home regularly. Table 5.4 lists the migrants' uses of their incomes. Migrants' buying power is rather limited. Most respondents stated that they were not earning enough money to save anything or to send any substantial amount back to their villages. As the follow-up interviews will demonstrate later, many migrants with the *intention* of remitting money or food in the off-season had not actually sent anything by the time the rains came. The concept of investment should be interpreted broadly; here it pertains to

TABLE 5.4   Uses of Migrant's Earnings

Uses of Migrant's Earnings	n	%
Satisfies basic needs only	77	57.9%
Satisfies basic needs and remits some	45	33.8%
Needs + agricultural investment	1	0.8%
Needs + remit + save	3	2.3%
Needs + remit + investment in clothing	1	0.8%
Needs + taxes	1	0.8%
Missing or not applicable	5	3.8%
	133	100.0%

investment in agricultural materials in the migrant's own fields, meaning buying chemical fertilizer, seeds, equipment, or similar items.

I was interested in finding out whether the male migrants' spouses and children work. I wanted to explore questions raised earlier about the ability of husband-and-wife pairs to exploit their own (and their children's) labor in the city. Displaying a lack of knowledge that was often striking, most of the men interviewed were not aware that their wives had any cash income at all. This response is one that most outside observers would find highly suspect. Almost 60 percent of the male migrants said that they did not believe their wives and children worked. My speculation is that the men I interviewed were ignorant that their wives were involved in the cash economy because the women were keeping such activities quiet in the hope that they could hold onto their earnings. Such underreporting of women's wages, particularly in informal economy activities, is very common, especially when national income estimates are based on the use of male proxies (Anker 1989; Goldschmidt-Clermont 1994). Additionally, women's secrecy about their earnings is likely to be justified out of a fear that their husbands will steal them.

Contacts Between Circular
Migrants and Rural Villages

Given that many migrants' choices of where to migrate were made partially for reasons of proximity, migrants depend on social contacts to assist them in finding work. That so many moves were undertaken with families' blessings indicates that the strength of ties to the village are apparently very great. All these imply that the level of contact between the city and the village will be substantial. Maradi is in fact a beehive of social activity, with ties extending across the city and between the city and the surrounding countryside. A large percentage of seasonal migrants maintain close contact with their villages through visits from family and friends and through return trips by the migrants themselves.

Overwhelmingly, migrants maintain contact with their family and friends back in the village. Over 95 percent replied that they stay in touch with village life through various means. The most common form of contact between respondents and their villages of origin comes through visitation by the migrants' family or friends, usually prompted by the twice-weekly market. Over 87 percent reported receiving visits from friends and family in the village during the preceding year, and almost 35 percent also made return trips themselves.

Much of the traffic in visits between village and city occurs with the migrant's own extended family. In all, 75.2 percent said that they received family visits. Answers seem to depend on distance and travel cost,

TABLE 5.5   Forms of Contact Between Urban Migrants and Home Villages

Forms of Contact	n	%
Visits from villagers	53	39.8%
Visits *and* return trips by respondent	46	34.6%
Visits *and* letters	9	6.8%
Return trips by respondent	8	6.0%
Visits, letters, *and* return trips	8	6.0%
Letters only	2	1.5%
No contact	3	2.3%
No response	4	3.0%
	133	100.0%

with farther villages sending fewer people to Maradi's market. Over half of the negative responses belong to migrants from Dakoro *arrondissement*, in the northern reaches of Maradi Department. Sending and receiving news in letters is relatively uncommon, but this is not surprising given the low level of literacy among the respondents.

Roughly half of the migrants reported sending cash gifts back to the village, though the size of these gifts is sometimes quite small. Much of this remittance money is used to purchase food and other necessities. Table 5.5 presents responses regarding the forms of contact between urban migrants and home villages.

Social conventions dictate that visitors from the village be provided with whatever they need for their stay in the city. Visitors are often given pocket money, clothing, and food. Social and cultural norms dictate that travel expenses, including the cost of their return journey to the village (*kud'in mota*, "car money") are paid as well. Table 5.6 lists the items migrants reported providing visitors from the village (multiple responses were possible).

As though in compensation for the largesse of their hosts, visitors often bring gifts themselves. These return gifts, often of agricultural surpluses, can ease shortages during stressful times. Some foods offered to migrants by their visitors include rice, groundnuts, cowpeas, dried peppers, dried tomatoes, onions, and millet. Other gifts include money, condiments, *gris-gris* (medicinal charms), and sometimes livestock. In all, 35.3 percent replied that they received gifts from visitors from their village.

The rationale for asking the migrant whether he or she has a family house in the village was to see how rooted the migrant still is there. In all, 81.2 percent replied that they still have a house in their village of origin, while 18.0 percent replied that they did not. Clearly the extended family's role in facilitating circular migration influences the answers to this question. In addition, sometimes migrants, especially very successful

TABLE 5.6   What the Migrant Provides Visitors

Money (unspecified amount)	88	66.2%
1000 or less	19	14.3%
1001 to 3000	4	3.0%
3001 to 10000	4	3.0%
Car money only	17	12.8%
Clothing	15	11.3%
Shoes	1	0.8%
Soap	3	2.3%
Other	11	8.3%
Food	27	20.3%
General food	12	9.0%
Salt	8	6.0%
Other	7	5.3%
Miscellaneous responses	14	10.5%
Nothing or not much	8	6.0%
Unspecified	5	3.8%
Place to stay only	1	0.8%
Nobody has visited	32	24.1%

ones, who do not have a family home in the village or who wish to improve the one they have will finance the construction of a house while they are in the city. Only rarely will a migrant maintain ties to the village if he or she has no family there. The house is a family home. Less than 7 percent of the respondents with a house back in the village stated that it was *not* occupied by a family member.

One of the most remarkable aspects of the migrants' existence is their continued ability to use agricultural fields in their village of origin. Even though ensconced in an urban labor market, in the rainy season migrants still leave their jobs to go home and plant, and this enhances their food security. The questions about agricultural fields were intended to probe the migrants' use of land in the village. In all, 88 percent replied that they had the use of agricultural fields in their home village.

The land tenure system is based on extended family control of plots and ensures that migrants continue to have access to land even when they are living in the city for most of the year. Many migrants maintain their contact with the village by farming village land themselves in the rainy season. Over half of the respondents said that family members farmed the land. Only 4.5 percent stated that paid labor, neighbors, or strangers cultivate the family's fields. Table 5.7 lists the family members responsible for farming the land in the rainy season (multiple responses were possible). In effect, agriculture plays a major role in maintaining circularity between migrants in the city and the sending places.

TABLE 5.7 Which Family Member Farms Land in Rainy Season

Which Family Member Farms Land	n	%
Self	49	41.9%
Family any response	59	50.4%
Whole family	24	20.5%
Any brother	16	13.7%
Big brother	8	6.8%
Father	3	2.6%

Open-ended questions were asked to determine the amount and use of remittances to the village home. Overall, 52.6 percent of migrants responded that they sent items back to their village, including money, foods, animals, seed, clothing, and fertilizer. Distinctions were made in the survey between *sending* items back to the village and *taking* them back personally, normally when the migrant returns to plant.[12] Table 5.8 lists the size and frequency of the remittances reported.

For those sending money home, amounts ranged widely but still were mainly modest. Payments made more often during the early part of the dry season tended to be larger. The mean size of trimestrial payments was 4,667 FCFA, the mean size of monthly payments was 3,750 FCFA. the mean size of weekly payments was 838 FCFA. To put the remittance amounts into perspective, a sack of millet costs between 5,000 and 10,000 FCFA. This means that perhaps 25–50 percent of the individual money gifts sent back to the village could be considered large enough to make a significant food purchase for the dry season. A remittance of 5,000 or 10,000 FCFA represents a sizable income boost for rural villages. Noncash remittances include soap, clothing, and fertilizer.

TABLE 5.8 Size and Frequency of Remittances

One payment per season	46	34.6%
Less than 1000	7	5.3%
1000 to 5000	15	11.3%
5000 to 10000	16	12%
10000 to 30000	8	6%
More frequent payments	29	21.8%
Trimestrial payment	3	2.3%
Monthly payment	2	1.5%
Weekly payment	2	1.5%
Money unspecified amount	7	5.3%
Clothing	15	11.3%
Non-cash gifts (total)	36	27.0%
Nothing remitted	60	45.1%

TABLE 5.9   Method of Transferring Cash Back to the Village

Method of Transferring Cash	n	%
Hand-delivered (total responses)	61	45.9%
Hand-delivered by respondent	18	13.5%
Hand-delivered by friends	7	5.3%
Hand-delivered at market	2	1.5%
Hand-delivered by children	2	1.5%
Hand-delivered by wife	1	0.8%
Post office	1	0.8%
Nothing remitted	60	45.1%

Specifying the mode of transferring cash to the village (see Table 5.9) sheds light on the mechanics of social networks. Wiring money through the post office, a practice known to exist in parts of Africa, is rather uncommon in Maradi, either because there are few post office facilities in rural areas or because of a growing fear of loss or theft. The most reliable method of transfer is hand-delivery by friends or family.

When asked if they knew what the remittance money they sent back to the village was used for, many respondents shrugged and replied that it was used for whatever needs the recipients had. When pressed, many (22.6 percent) said that they thought the money was used for food purchases or for miscellaneous necessities. Far less frequently, they replied that it was used for investments in agriculture (such as chemical fertilizer or seeds).

Questions about what season the migrants returned to the village were asked to probe the effect distance plays in the timing of return visits or stays. Overall, 74.4 percent replied that they had returned to their village in the preceding twelve months. As with family visits, most of those who had not returned home came from more distant villages, mostly from either Dakoro or Mayahi *arrondissement*. Proximity plays a role in determining whether a migrant makes return trips to the village during the dry season, but a majority (60.9 percent) go home to plant after the rains start. As for their motive for returning, 65.4 percent said they returned to do agricultural work, and 32.3 percent cited family reasons, including ceremonies. Circular migrants continue to return home year after year because of their extended family's need for on-farm labor and the migrant's need for food.

Again, as with the questions pertaining to the money sent back to the village, it is difficult to determine exactly how critical the gifts of cash and food are to the recipients. But it is clear that offering them fulfills both economic and social needs. In all, 38.3 percent of the migrants took money home with them when they returned to their village. Of the migrants who had returned in the preceding twelve months, over 45 percent took money back with them when they returned. Other gifts included agricultural materials such as seed and fertilizer. Given this high level of contact between migrants in the city and the extended family in the village, I was interested in exploring whether there was a societal stigma attached to people who permanently cut their ties with their village of origin. I had heard that some wealthier migrants resist contact with the village, wary of obligations to support impoverished family members and acquaintances. I found the rare cases of cut ties to be related almost invariably to specific life-course events, and specifically to divorce or the death of family members. One male migrant, Labdou Ibro, who now makes fencing in the post-harvest season, had worked through the 1960s as a docker in Nigeria. He had remained away from Niger and out of contact for many years. While he was away, his parents as well as all four of his siblings died. Understandably, he did not feel a need to keep in contact with the village.

As this section has illustrated, Maradi migrants maintain intricate and durable ties to their village homes, even after they have moved to the city. In most cases, they make frequent contact with the village, sending money, food, and agricultural materials, including fertilizer and seeds. In return, they receive visits from friends and family in the village and sometimes gifts of food. The depth of these ties raises implications for the migrants' existence. Are the migrants city residents, or are they in fact caught between the city and the country? In this case, the rural-urban distinction is not very helpful.

Continuity of Mobility Practices

Questions about migrants' social life in the city were intended to explore whether social organizations of any type have emerged to take the place of the family support that migrants enjoyed in their villages prior to leaving. On the whole, the doors of formal organizations in the city are not open to migrants. In all, 85.7 percent replied that they did not belong to any formal organizations in the city; 13.5 percent said they did belong to formal organizations in Maradi, including migrant organizations, a *caisse populaire* (community bank), a Koranic study group, a political party, a karate club, a local Red Cross chapter, the local radio-TV office, and an association of practitioners of traditional medicine. One former wrestler belonged to a group called *Dan Banga* (Son of the Drum), a vigilante organization that patrols the streets of Maradi at night looking for thieves.[13]

Questions about the migrant's future outlook in the city were devised to probe whether migrants truly feel at home in Maradi and whether they intend to remain in the city. Results were heavily on the side of residential impermanence. Overall, 85 percent replied that they only considered Maradi a temporary home. The impression from the migrants is that they will stay in Maradi unless or until something better comes along, and this makes their moves persistently temporary. Maradi's migrants are held in place by a balance between opportunities in the city and ties to the home place. As long as both needs are being satisfied, they will continue to stay. As soon as one element is no longer productive, the migrant will leave. Their commitment to the specific place where they live, if such a place is equated with a physical site and location, is therefore low. By focusing on the non-formally educated migrants in the informal sector, perhaps I biased my study away from those in more permanent jobs or those who have made a deeper commitment to rooting themselves in the city. Even so, the migrants I interviewed represent far greater numbers than the category of permanently displaced, professional, and formally educated civil servants.[14]

In their daily habits, migrants have actually become urbanized in only a few ways. Most are not connected to any modern social infrastructure. They live in enclaves within the city and rarely have much use for such elements of the modern infrastructure as movies or telephones. They intend to be in the city only temporarily, and they rarely join formal organizations in the city. They are like many first-generation immigrants: Culturally and psychologically, they still live in their former homes.

Questions about their levels of satisfaction with their stay in Maradi were intended to provide further insight into the temporary state of the migrants' existence in the city. Overall, most of the migrants (82 percent) stated that they are content with their stay. They give various reasons. For the most part, satisfaction seems to be based on material conditions. This was often expressed in thanks for reliable supplies of food and water.

Of the 10.5 percent who were not satisfied, several stated that there was currently "not enough of anything" or that their situation was not good for them because there was no work. Two women replied that hunger had forced them out of their villages and that they were not in a position to do anything except ensure their children's survival. These women were among the few who felt that migration had been forced on them and was not a choice.

A second "outlook question," about whether the migrants would advise others to come to Maradi, was asked in order to gauge whether their experience has recommended itself to them. In their responses, which are not overwhelmingly positive, only 48.1 percent replied that they would recommend migrating to Maradi to prospective migrants. This response might actually have very little to do with how positive the migrants' stay has been; rather, it could reflect a desire to preserve the good qualities of the city. Many respondents reasoned that there were already enough people in Maradi and that if they recommended Maradi and many others came, the city would be inundated with migrants and the quality of life for all residents would suffer. One man said that he could barely satisfy his own needs, so he could not recommend migration to others. Others said that they had had to migrate or that they were not in a position to recommend it, since they themselves had suffered.

Asked if they preferred the city or the village, many migrants replied that they would clearly prefer to be back in the village with their parents, and in some cases with their own spouses and children. Some regarded my question as ridiculous, feeling that they should not contemplate a hypothetical situation over which they have no control. In all, 46.6 percent of respondents replied that they prefer the city, and 51.1 percent replied that they prefer the village.

These sentiments seem to stem less from the changes in the region than from their own situations. Many saw the changes occurring in the region as inevitable. Moustapha Kirki, age twenty-seven, responded that tradition was what made people migrate: "The land is rich, but even so, people are obligated to migrate south from the northern places." Abdourrahaman Idi, age twenty-two, said, "One cannot just stay at home." These migrants had left the village, but they would prefer that the city *not* be inundated with people in similar straits. Perhaps their wistfulness is brought on by a feeling of having left behind people they are obligated to help. They *intended* to support their families, but obviously they cannot do as much as they would like, given their predicament in the city. For the migrants, this is a very touchy subject.

A question about the "solutions to migration" was intended to elicit the hypothetical qualities about the villages of origin that would have to change to prompt the migrant to return there. This sometimes sparked discussions regarding "the problem of migration" and the respondents' perceived solutions to it. Responses are listed in Table 5.10. Many of the responses reflect an awareness of changes in the land that have made food production more precarious. More often on migrants' minds, though, was the issue of work. Abdou Illiassou, age sixty, replied that "it's true that there isn't enough land, but it's more pressing to us that there isn't any work to do in the village." Amadou Souley, age fifty-five, stated the dynamic more bluntly: "When food runs out, people migrate: It's simple." Hawa Hada, age forty-two, felt that the need to move was inevitable: "People get upset with conditions in the bush and feel the need to leave." Yacouba Issa, age thirty-two, maintained that the problem is food, saying, "You know as well as I what the problem is!" It is striking how many of the observations about life in the village stem from the perceived lack of something, which seems at times to become apparent only after the migrant has left leaves the village.

TABLE 5.10   The Perceived Causes of Migration

Food	56	42.1%
Lack of food	44	33.8%
Not enough millet growing	4	3.0%
Other	8	6.0%
Work	42	31.6%
Lack of money	14	10.5%
Lack of work	11	8.3%
Other	17	12.8%
Land	17	12.8%
Lack of land	4	3.0%
Other	13	9.8%
Personal reasons	15	11.3%
Marital problems	3	2.3%
Suffering	3	2.3%
Other	9	6.8%
Miscellaneous reasons	13	9.8%
Climate	9	6.8%
Drought	8	6.0%
If rain, he could do off-season gardens	1	0.8%
Agriculture	7	5.3%
Lack of animals	1	0.8%
Lack of water	2	1.5%

A hypothetical question about whether the migrant would return to the village *if* there were paying work there elicited mixed responses. In all, 69.2 percent responded that they would return if they could earn cash in the village, and 23.3 percent replied that they would remain in the city even if they could go home. About 2 percent replied that they would return if certain other conditions were met. The question split the group into those who see the benefits of the migration for themselves and those who migrate more or less out of desperation, driven by material shortages such as seasonal shortfalls. Some responded that they liked the city and would migrate anyway. Others replied that they would not return to the village even if there were food. One migrant voiced a sentiment with which I feel many other migrants would agree: He preferred to have both the city *and* the village. Some people made their a conditional reply, saying that they would go home "if all went well," "if there were good work," or "if there were a project in the village."

This investigation into this context for urbanization has revealed considerable diversity within the population identified as migrants. In characterizing the migrants, it is difficult to determine exactly how they place themselves between the city and the countryside or how marginalized they actually are. This study has sought to discover how integral migrants are to the life of the city. In effect, they *are* the life of the city; it should not be forgotten that their presence in Maradi is good for the city too. In fact, 85 percent of the migrants consider their stay in Maradi to be temporary and plan eventually to return to the village. At the hub of a complicated network of social ties, the migrants have secured for themselves a more advantageous locational position than they ever could have by remaining in the village. In this way they have capitalized on their circumstances, becoming, in reality, urban residents with enough flexibility in their work schedule to permit them to return home to grow food. With its locational advantages, this bilocal residence is noteworthy, as is the resemblance between the urban squatter settlement and the rural village. This raises the question of where the migrants actually live, and where "home" is considered to be.

Notes

1. *Masu digga* has no precise derivation other than that denoting temporary migrants. The verb *digga* in Hausa means "dripping" or "dribbling out in small quantities" (Newman and Newman 1977).

2. Market days are scheduled so that one can progress in a specific direction and sell one's wares at different markets on consecutive days.

3. The only existing study done on Maradi migrants, by Raynaut et al. (GRID 1990) pointed out such segregation of migrant neighborhoods within the city. These neighborhoods are typically later-settled *quartiers* such as Zaria, Ali Dan Sofo, and Sabon Carré, which were built and filled during the droughts of the 1970s and 1980s.

4. My interpreter, Kanta Wakasso, played an important role in finding migrants and then in assisting with the survey-interviews. The fact that survey forms were written in French and translated by Kanta into Hausa during the interviews created a definite language barrier. Issues of potential sources of bias and error are addressed as follows: *Sample size* is considered a nonobservational error that affects coverage. *Sampling errors* reflect the discrepancy between those in the total population who did not fall in the sample and those surveyed. Because of the "snowball" approach I used and the lack of systematic sampling techniques, there is no statistical way to verify how representative the 133 people I interviewed are. Increasing the sample size would reduce the error. In a city of perhaps 165,000, containing thousands of seasonal or temporary migrants, polling a sufficiently large randomly or systematically selected sample would be a costly and laborious undertaking. I do not feel that the present study warrants such an approach.

Nonresponse or nonobservational errors stem from people who belong in a statistical sample but did not respond or were otherwise not measured. Kanta and I did not feel this was a critical issue. We did encounter groups in some *quartiers* who mistrusted us initially because they thought we were collecting names for tax purposes. This was reflected in the difficulty Kanta and I had at times in finding people who would admit they were born in villages outside of Maradi. Sometimes migrants are reticent about the topic of residence, apparently because they are avoiding their taxes back in the village. Mostly though, we were warmly welcomed and never lacked eager participants.

Observational errors often stem from the practice in survey research of employing trained interviewers to administer questionnaires and to navigate through often complicated sequences and skip patterns. Such errors were not totally absent in the present study, but I believe their effects were minimal because I participated directly in every interview, and the questionnaire (which was developed by Mounkaïla Harouna, a Nigérien geography student) was used rather loosely. I scribbled notes freely on the survey forms, often wrote down my impressions at length after interviews, and also tape-recorded some survey-interviews, though this method was found to be intrusive. The short format (thirty to sixty minutes) obviously did not capture everything. Personal biographical details were lost that would have found a place in a life-history collection lasting hours or a participant observation extending over days, weeks, or months. But for the purposes of the study, which was to become acquainted with the migrants as a population and also as people, the survey-interview format met the needs of the research agenda.

5. According to Zachariah et al.'s 1980 study of migration in anglophone West Africa, the ratio of male to female migrants in Ghana, Sierra Leone, and the Gambia is closer to 2 to 1. Perhaps the more skewed ratio in Maradi is due to the effect of Islam. Some of the explanation may also perhaps be due to biases inherent in the research design.

6. In the relevant age groups for migration, 86.5 percent of men and 97.3 percent of women in the twenty-five-to-twenty-nine age groups were currently married. In the thirty-to-thirty-four age group, 93.9 percent of males and 95.7 percent of females were found to be currently married (UN 1979).

7. There is of course no reasonable way to quantify the exact percentage of those in distress.

8. The ages listed for the migrants were those reported in the fall months of 1995, when the survey-interviews were conducted.

9. Generally bush taxis and buses cost about 100 FCFA (US$0.20) per ten kilometers traveled.

10. One must bear in mind that cousins or other relatives not literally in the migrant's immediate family were also considered "big brothers."

11. There are good reasons to think it would not be this high throughout the year, especially in the rainy season.

12. These two English words both translate to the Hausa word *aika*, so confusion resulted from time to time.

13. These vigilante organizations must do an effective job of keeping the streets safe at night. Compared with cities in Nigeria, even in the Moslem north, Maradi is exceptionally safe.

14. According to Painter's (1987) estimate, there were between 100,000 and 350,000 seasonal circular migrants in Niger in the mid–1980s. This contrasts with the estimated 27,000 civil servants and teachers prior to the governmental crisis in the early 1990s (Charlick 1991).

6

Migrants' Livelihoods

Malam Harouna Issa, the *marabout* whom we met earlier, happily agreed to meet for a follow-up interview six months after my initial visit. I organized the meeting in Maradaoua *quartier* in Maradi-*ville* to discuss some long-standing questions that had arisen during the individual interviews. I wanted to ask about migrants' obligations to send money back to their villages, behavior during periods of drought, and some underlying causes of periodic crises in the region.

The dynamics of the group discussion presented themselves to me immediately. All present deferred to the *marabout*, who clearly relished the opportunity to talk to me again. When I tried to pose questions to others, nervous laughter and hesitation would follow, until Malam Harouna broke in to respond. Later this tension eased a bit, and others began to speak for themselves more. For the first time as an interviewer and an observer, I felt that my clipboard and written notes were out of place. The weight of the group bore against me. Even with my friend Kanta next to me to translate, I felt cornered. My conversation with the *marabout* was perhaps the most revealing one I had in Niger during my research stint.

This encounter, which I will recount in detail later in the chapter, underscores the point that migrants play an active role in engineering their own circumstances. The objective of Chapter 5 was to look at Maradi's migrants as a sample of a moving population; the objective of this chapter is to see the migrants on their own terms. The migrants emerge as proactive decisionmakers when their economic strategies are explored and the population is broken into more meaningful social units. Changes in the contexts of migration during the 1990s amplify the prevalence of certain traditional occupations, and this process reveals how fluid labor has become as a medium between rural and urban. What circular migrants do naturally lends itself to analysis because, as we will see, the occupations have their own genealogies.

TABLE 6.1   Residence in the City and Estimated Daily Earnings

Profile Group	n	Average Number of Years (seasons) in Maradi	Average Earnings per Day (est.) FCFA	in $US
Guards	14	7.8	751	1.50
Post-harvest activities	17	7.1	510	1.02
Services	26	8.4	1087	2.17
Crafts/semi-skilled	14	13.6	1461	2.92
Women's occupations	14	10.6	698	1.40
Market/sales	21	6.5	938	1.88
Farming	7	8.2	131	0.26
Community leaders	8	13.1	340	0.68
Nonemployed	12	9.5	42	0.08
Total/mean	133	9.0	763	1.53

Rural to Urban Transformations

The informal sector—defined earlier as any economic activity not recorded in national income accounts—*is* the economy of Maradi. Its development is itself evidence of a transformation whereby off-season rural occupations have moved to urban centers and have been absorbed into the economy. Table 6.1 presents nine occupational categories. It lists the average length of stay in Maradi of the migrants engaging in those occupations and the migrants' estimated earnings in both FCFA and U.S. dollars. The table shows the length of migrants' residence in Maradi rather than their age because the length of stay gives a better idea of the migration capital that migrants have accumulated. It should be noted that a "year" in Maradi is equivalent to one dry season. Earnings are based on respondents' own estimates and include only cash, not food or gifts. This is evidenced in the very low earnings of some groups, notably community leaders, farmers, and the nonemployed. The average daily earnings are approximately US$1.50.

Fourteen of those interviewed were *masu gardawa* (guards). Guarding is a relatively new occupation, dating to colonial times. Guards are responsible for security at larger houses in town, especially at merchants' and expatriates' villas. Guards are often pastoral people, frequently Tuareg or Fulani who have lost their herds or who have migrated to the city to earn cash. They carry swords and have a well-earned reputation for ferocity. Their median reported income of about 750 FCFA per day is not large, but the position of guard is coveted because of its stability compared with other jobs in the city. Guards may work either during the day or at night, and they are individually hired for qualities of independence and willingness to work alone at night. The average age of those interviewed was 44.6. It is a relatively high-paying, steady occupation requiring experience and social capital. Guards need to know and talk with one another, which allows a developed network to exist on streets containing large houses. Crime is not a huge concern in Maradi—it is not as serious as in many African cities—but petty thefts do occur.[1] There was a murder in Maradi a few years ago that was associated with the political parties. Traffic in weapons and drugs is for the most part low, and the city is tranquil at night for most of the year.

Tidjani Dama, twenty-seven, was born in Kadani in Dakoro *arrondissement*, which lies in the Tarka Valley perhaps 120 kilometers north of Maradi. He is from a family of Bororo cattle herders. He left his family first in 1990 at age of twenty-two to move to Maradi, having already visited the city many times. Bright and outgoing, he fell in with the Peace Corps volunteers and soon found himself hired by one to work as a guard. His ease in the city is shared by other pastoralists, despite what is often overt discrimination by the sedentary majority. Tidjani's wife, Aisha, lives with him in the Peace Corps compound in Maradi, though she has been known to migrate from time to time. They have no children.

Tidjani's older brother, Boubé Dama, thirty-seven, followed him into the Peace Corps network. Boubé has actually been a migrant longer, since the age of twenty-eight, working in Kano and Kaduna as a guard and selling traditional medicine. Boubé's motivation for his first departure was the loss of family cattle herds to drought. A calm and peaceful man, Boubé looks dreamily at pictures of North American elk and deer, which he calls *naman daji* (bush meat), in the *National Geographic* magazines the Peace Corps volunteers share with him.

Both Boubé and Tidjani spend much of their free time embroidering purses and clothing, which they then sell to Peace Corps volunteers as gifts for their families and friends. Boubé is a member of a crafts *caisse* (revolving credit association), which loaned him money to buy thread and other supplies to get his business started. Like Tidjani, Boubé makes visits home whenever he can. Both visit their extended family a few times per year and entertain their family when they come to Maradi. When Boubé visits home, he usually takes cash (5,000–10,000 FCFA) and a sack of rice as presents. The last time I saw him, Tidjani had just invested in a very expensive and elaborate camel saddle, which he intended to resell. The biggest hardship experienced by the brothers, though neither of them would dwell on this, is the stream of visitors from the countryside that they are obliged to entertain. People passing through Maradi en route to Nigeria drop by for a meal or to spend the night, and this creates something of a financial burden for Tidjani and Boubé, especially when their wives are away and they have to purchase large quantities of food for the guests. Neither of the brothers can cook.

These former nomads have used mobility to make a successful transformation to a new lifestyle, redirecting their old customary routes to include the markets and the bright lights of the city. But having lost their grazing lands and herds, they are unavoidably wistful about their previous lifestyle. As modern migrants, they have been forced to shed some of the trappings of the pastoral existence of great desert clans, yet they still find the urban experience liberating. This hopefulness can also be seen in Balkissa, Boubé's wife.

Seventeen of those interviewed worked in post-harvest activities, including making ropes, fences, and roofs, using millet stalks and purchased or scavenged doum palms. Post-harvest activities are new as a migrant occupation. Traditionally, such activities were unmonetized and performed in the village after the end of the rains in October and November. Today they are cash-generating activities for the urban economy, conducted by the *masu digga*. Their products are sold at the biweekly market to people rebuilding their houses and compounds after the assault of the annual rains has ended. The individuals I interviewed were men arriving from the same area, north of Dakoro, who were living together in the city. The group displays a tight organization, preparing and eating food. Their median income of about 510 FCFA per day should be considered an extremely loose estimate, because the post-harvest season forms the busiest period for fence and rope sales. The average age of this occupational group is 33.2 (Photo 6.1).

Daouda Issaka, twenty-eight, is from Goukou in Dakoro *arrondissement*, also in the Sahelian agro-ecological zone. Daouda was born the eighth child of a family of more than thirty (from more than one wife), living in "very much a subsistence situation," where sometimes he ate and sometimes he didn't. Asked if there have been changes in his village since his childhood, he described a general decline—there are fewer trees in the village than before—but he has no rosy reminiscences of plenty. "The land was always bad," he said. Daouda's father was also a seasonal migrant. I asked Daouda if he thought that migration had any benefit for the village of Goukou. He replied that if there were any benefits, they were not very visible. By migrating to the city in the dry season, Daouda removes himself from the need to be fed from his parents' granary. His absence saves his family's food, but he sees little profit in it besides that.

Daouda represents the new generation of migrants, who are temporarily displaced and doing in the city what was until recently essentially village work. At the time of the first interview, Daouda had only been in

Photo 6.1   This group of men moved to the city temporarily after the end of the rainy season to make rope and fencing materials for cash. Their work represents examples of village activities that have been transferred to the city's cash economy in recent years.

Maradi for six days. He left his wife and their eight children behind in the village to travel to Maradi, where he joined a group of men from the same area who make fencing out of millet stalks and sell it behind the Maradi *tasha*. In previous years, Daouda migrated to the Nigerian towns of Fantua, Jibiya, and Katsina, working as a head porter, an apprentice mason, and a water carrier. In this season, if he keeps at it all day, he can make about 5,000 FCFA per week, and he has hopes of sending some of his earnings back to his family in Goukou soon.

Adamou Gouda, thirty-five, is from Rijia Makaou, another Hausa village in the later-settled midsection of Dakoro *arrondissement*, and his first migration occurred in 1980 at the age of twenty. His most recent migration was made with his family on foot, walking from the small village of Dara to Maradi by way of Gangara. He arrived six days prior to the interview. At the time of the interview he had not sold any fence and so could not estimate his prospects for the year in the city. On my follow-up visit to the fence makers on their patch of land near the road to Madarounfa in May 1996, they had revised downward their high hopes from the previous fall about making a profit from their work in the city. Adamou said laughingly that he would go home with *kadago* (a twenty-five-franc coin worth a U.S. nickel) in his pocket.

Why weren't the fence makers able to save more? The cost of living in the city had risen since they arrived the previous fall, and the cost of *fura* (millet porridge) had gone up from 150 to 250 FCFA per bowl. Neither Daouda Issaka nor Adamou Gouda had sent much home, other than the news they were all right. Neither of these men would be here if they could stay in the village and support their families. When asked what the solution was, Ilia replied that there needed to be more rain in his village. The lure of the city does not mean much to them, since they spend most of the small wages they make on buying food. They cannot afford to visit prostitutes.

The post-harvest migrants' predicament is plain. They come from marginal places in the north, where food security is maintained through seasonal absence. For them, their circumstances—changes in the land, too many people, and the growing need for cash—are all wrapped up together, but the bottom line is the lack of food. They are distressed people from distressed places, but even so, they are likely to be doing better than their families back home.

The category of services is rather a catch-all, covering a variety of entry-level, unskilled occupations and including apprentices, a baker, a barber, a brick maker, a laborer, a launderer, a porter, a singer, and various repairmen. Such workers gain access to opportunities through social networks, apprenticing themselves to an established person or setting out on their own. Their median income is about 725 FCFA per day, and the average age is 33.2. Many service workers are wading into the city for the first time. It should be remembered that among the service workers, 56.4 percent relied on family or friends for support after arriving in the city.

Inoussi El-Hassan, forty-three, works as a driver for a development organization in Maradi. His higher earnings make him an exceptional representative of the service category. Inoussi is from Madarounfa in the extreme south of Niger; compared to many of the other migrants I interviewed, he has indeed traveled a distance from bare subsistence living. Inoussi attended primary school for six years, he lives in a rented mud house, and he pays other men to farm his land. Unlike many service workers who cannot claim to be so independent, his salary is not tied to the vicissitudes of the urban market. Inoussi went on his first migration at age eighteen in 1970, traveling first to Niamey, where he learned to drive, then north to Arlit to work in the uranium mines for two and a half years. Later he worked in Tahoua and Zinder before settling in Maradi last year, to work for Stichting Nederlandse Vrijwilligers (SNV), the Dutch aid organization. He makes 55,000 FCFA per month. With the economies in Nigeria and Benin showing strains and the ripple effect threatening to wash Niger away, development aid is one of the few viable large enterprises in Niger. Being a driver for an international organization like SNV or CARE is every urban taxi driver's dream.

Inoussi's wife and children live in Maradi, and he has stable year-round work, but he still maintains close contact with his home village, receiving visits from family and making visits himself, even if only for the weekend. After the harvest, he will often receive a sack of millet from his family, and to reciprocate he will give his family clothing and money. The cash gifts he has provided his family in the past have bought food to cover food deficits, but he is quick to point out that sometimes *he* asks his family for some money. He supports an elderly parent in ill health who has moved to Maradi. Simply put, Inoussi's attitude about migration is: "If there were food and water in the villages, then people wouldn't leave." Yet he is a shining example of someone who has made something of himself and done well by going away.

Another migrant in the service category, forty-four-year-old Souley Maazou works as an itinerant laborer in Zaria *quartier*. A stooped man who speaks quickly and excitedly, Souley comes from Dakoro Maïjéma in Dakoro *arrondissement*. He left on his first migration when he was eighteen, walking to Kornaka and then to Maradi over the course of four days, along with his family. Souley has worked as a guard since 1969 in numerous West African cities, including Lagos, Jos, and Kano. He does not remember knowing anyone when he arrived in Maradi, and only with the help of divine force did he find food and a place to stay. Now he and his family live in a straw house in Zaria, along with many other migrants. His wife pounds millet, and his three children work in Maradi. Souley has made many acquaintances here over the fifteen years he has lived in town. He is typical in the level of quiet dignity he displays in his daily doings.

Souley still returns to his village to plant during the rainy season. Asked if he is satisfied with his stay, he defers because he is obliged to stay in Maradi. He would prefer to stay in his village and would not advise others to come, because he himself does not have enough, but staying in the village is not practical. His solution to migration is "more food." He states that there is plenty of land in his village, which lies in the north, but that there is not enough food there.

Fourteen migrants were identified in the crafts/semiskilled category, which, as the name implies, requires a modicum of skills. This category is based in part on traditional guilds for independent artisanal activities. The small operations that employ these workers are based on apprenticeship, which relies on implicit agreements between the *patron* and the apprentice to govern work arrangements. The apprentice receives irregular earnings, but they are enough to enable him to keep working, to purchase food and lodgings, and to have some pocket money. In return, the apprentice provides his labor. Such arrangements are often marked by very intense competition among apprentices, and of course the work must necessarily be seasonal (from October to May), allowing time off for planting back home. Microeconomic activities are characterized by traditional crafts guilds (*sana'a*) in metalsmithing, dyeing, weaving, tanning, masonry, butchering, and barbering, which are often passed from father to son (Hugon 1980). Since the drought years of the 1970s, microenterprises in Maradi have become essential to the town's absorption of immigrants (Grégoire 1992) and also to the city's economy.

The crafts/semiskilled category as defined here includes practitioners of traditional medicine, a leather sandal maker, a wooden bed maker, a hatmaker, a *marok'ai* (praise-singers, known in French as *griots*), a sword maker, a carpenter, a tailor, a butcher, and an itinerant snake charmer. The median income of this group is distorted seasonally by high wages in the post-harvest season, when it averages an estimated 1,500 FCFA per day. The average age is 41.9.

Abou Hamza, the sword maker introduced in Chapter 4, is sixty-five. He comes from Azarou near Madaoua, in the Sahelo-Sudanian zone. He lives in Soura Bildi *quartier*, in a mud-brick house that he owns, in a shady compound. Abou left the village for the first time in 1971 at age forty-one, coming directly to Maradi with no other intermediate destinations. Such focus is unusual, and it surprised me. I asked why he had chosen Maradi. "It's all I heard: It's the economic capital of Niger, and it has rich people, and there wasn't a chance I would suffer." He shares his house with two wives and four children. The advantage of his livelihood is that it can be practiced anywhere, and he is almost guaranteed a market wherever he may choose to go. Abou sells his wares at the market and around town, earning an estimated 1,500-2,000 FCFA per day. His contacts with his village of origin are still strong; just a few days prior to our interview he had hosted a visitor from Azarou. He receives members of his family of origin as well, to whom he gives pocket money. Sometimes he receives some money or a sack of millet from the village. His family of origin still farms in Azarou, but Abou no longer does, and he does not send as much money home as he would like. The last time he returned for a naming ceremony, he gave his family a 15,000 FCFA gift. Although he has definitively made the move away from Azarou to Maradi, he still calls the village his home.

I interviewed fourteen women in Maradi. Their occupations reflect, in a way, the limited opportunities for women in the city. Hausa cultural constraints can be seen in the fact that only postmenopausal women, divorcées, and prostitutes—known euphemistically as *karuwai* (free women, or courtesans)—were available for me to interview in Maradi. The group includes six women working in the preparation of various foods, including millet porridge, fried bean cakes, and cooked yams for snacks; a milk seller; a practitioner of traditional medicine; three prostitutes (prostitution is easily the most lucrative of women's occupations);[2] and two millet pounders.

Millet pounding for a cash wage has an unknown history; it is almost certainly not a traditional commercial activity in the city. Pounding is another example of a rural village activity that has been moved to the city and monetized. It probably reflects the needs of a growing number of women working outside the home, who no longer have time to cook. Millet pounders usually live in or near the compounds of their hosts and work for 2,500 FCFA per month. This wage frequently includes food and lodging, but still it is the lowest I encountered in my work with Maradi migrants. The median income of the women's occupations is about 375 FCFA per day, and the women's average age is 41.4.

Zalika Salifou, forty-five, is from Dan Bouzou in Mayahi arrondissement, a Sahelian village about ninety kilometers northeast of Maradi. She lives in the shantytown built on the old hippodrome in Zaria *quartier* and works as a prostitute. She smoked a cigarette during our interview, a habit women rarely indulge in in public in a Nigérien village. She was born the tenth child in a "very large family," and nine of her siblings are still alive. In fact this survival rate is unusual: Several migrants I interviewed said they were the only surviving offspring of some very large families. Zalika still receives visits from her brothers. Zalika reports that she had just barely enough to eat, and nothing more, when she was growing up. Her initial migration to Maradi coincided with the breakup of her marriage. She left Dan Bouzou without her two children and stayed in Maradi with a friend from the same region. She still goes back to the village to cultivate, and she sends money whenever she can to her children. Her husband has remarried and does not want her in the village, so she has to be secretive and stay with friends when she goes home.

Karima Abdou, fifty, is from Madata near Madaoua. She is divorced and has four children, who live with her in a cluster of millet-straw huts behind a roadside truck business. Karima pounds millet for a fixed monthly wage, earning approximately 2,500 FCFA per month. Her friend Hassana Idi, forty, is also from Madata and is also a millet pounder. Hassana left the village with her two children in 1991 because of famine, and she found work in Maradi. Her husband, who divorced her around that time, is currently a migrant too, though he lives elsewhere. Both Karima and Hassana maintain contact with the village through letters and visits, though no one is currently living in their houses in the village. Friends pass through and bring news from home. When Karima returns in the rainy season to plant, she takes one or two sacks of millet to use for seed.

Asked if they were satisfied with their stay, Karima and Hassana are unsure of how to answer. "It's a necessity," Karima says, caused by the lack of food.

When I returned in May to conduct a follow-up interview with these two women, Karima and Hassana were gone, and there was a different group of families staying in the cluster of millet-straw huts. "They lost their health and left several weeks ago," the new residents explained.

The migrants engaging in market/sales activities are dependent on the biweekly market or on itinerant and individual sales in Maradi. Six of the twenty-one interviewed identified their occupation as associated directly with the market. Five were *masu tebur*, that is, roadside vendors of cigarettes, gum, batteries, and other essentials. Other various vendors, of milk, peanuts, cigarettes, clothes, condiments, gas, and furniture, are included in this group as well. Marketing occupations are traditional in the sense that they preceded the colonial era as well as the distorting effects of the current political-economic crisis. Their median income is about 570 FCFA per day. The average age is 35.2.

Mani Bara, fifty, is from Tamaské in the Sahelian zone, about 265 kilometers to the northwest of Maradi. He sells metal kitchenware from Nigeria near the old Dakoro *tasha*. Although his father was not a migrant, Mani fits the description of a "classic" seasonal migrant, having worked as a porter in Birnin Konni, Sokoto, Kaduna, a city in Burkina, and Abidjan for many seasons before coming to Maradi, where the husband of his elder sister lived and took him in. He rents a mud-brick house, is married, and has six children. Mani sees people from the village from time to time at the market, but he does not receive much in the way of news from Tamaské anymore, perhaps because it is too far away and he has been gone for so long. Mani says he does not send his family any food or gifts, and he had not made the trip home in the preceding three years. His big brother still lives there, but they have fallen out of contact. Still, Mani sees his stay in Maradi as temporary and will return home in the future "if I get something."

Mani remembers the changes in the village since his childhood to be mostly positive ones. There have been improvements in agriculture, and now there is a middle school and a livestock veterinary center. He does not believe there is a land shortage in Tamaské. When I asked him if a lot of people left Tamaské during the last drought, he replied that many had gone to Nigeria; some had returned but others stayed away. Mani maintained that going to the city is the only way for a villager to become rich.

The farming occupation is represented by those who grow food year-round in the irrigated lowlands to the west of the city and by those who lack a secondary, nonagricultural occupation. Farming is a residual category, and the seven farmers interviewed were not necessarily recent ar-

Photo 6.2   A marabout *stands for a portrait against a millet-stalk fence.* Marabouts *are regarded in the city as keepers of the village tradition and can sometimes be vociferous defenders of traditional ways.*

rivals. Only 43 percent of them have been in Maradi five years or less, as opposed to 66.7 percent for those engaged in the market/sales occupations. The mean reported income is zero, with a median of 131 FCFA per day. The average age is 39.0.

Abdoullaye Karou, thirty-one, is originally from Alhourma Guidan Bounou in Mayahi, not far from Guidan Wari. His parents now live in Maradi too. Abdoullaye is married and has four children. He farms rented land outside of town, on which he grows market vegetables such as tomatoes and eggplant. He lacks a secondary nonagricultural occupation, possibly because he simply has not found one yet. He began migrating at age nine. When there was no crop, he left his village and came directly to Maradi. His older brother still lives in Alhourma Guidan Bounou and maintains the family fields. This coming summer, Abdoullaye will go back and help his brother. If he has the money, Abdoullaye will pay some men from the village to help him prepare the fields there.

Eight migrants fit the description of community leaders. This group includes several *marabout*s, the Islamic religious leaders (Photo 6.2), and one *animateur* (community organizer). *Marabout*s are supported by their neighborhood to teach the Koran to youngsters in outdoor street-corner classrooms. Owing to school strikes, a lack of money and educational materials, and reductions in stipends, the Nigérien public school system has slowly lost its national role. Koranic education has become a popular alternative for children. *Marabouts'* cash earnings are normally very low, but their occupation carries both cultural and religious significance and a considerable amount of social power. *Marabouts* are often long-time members of urban society and are often highly respected men. Because so many of them reported their incomes as zero, the median income is about 35 FCFA per day, with a mean of 340 FCFA per day. The average age is 43.4.

Moussa Diallo, twenty-seven, is from Guidan Dawo in Mayahi in the Sahelian zone, about fifty kilometers from Maradi. Guidan Dawo means "house of return" in Hausa. As a *marabout,* he has been acculturated in the practices of mobility, having studied the Koran and begged in many Nigérien and Nigerian cities since around the age of eighteen. He is a street pastor in Ali Dan Sofo *quartier*, tending mainly to the flock from his region in Mayahi who circulate between the village and the city. Still, he returns to Guidan Dawo every summer to plant and visit with his family. Moussa was the firstborn of nine in what he describes as a moderate-income family, "not rich, but my parents fed their children." His father was a seasonal migrant too. Moussa reports that the soil fertility has diminished in his village since his childhood, and people there now must apply manure on their fields for a good harvest. There are fewer trees and animals too. Moussa remembers twice in the village's history when many people left during droughts. But there is still available land in Guidan Dawo. I showed him the aerial photo from the 1950s for his village, and he dismissed the possibility of a connection between occupation of the land and the tendency to migrate. Moussa's feeling is that people are migrating not because of land shortage or low yields but because it is something they've always done, since there is no paying work in the village.

Yacouba Ibrahim, forty-nine, is from Gakoudi, also in Mayahi's Sahelian zone. He has a powerful and direct gaze and great warmth, so it is not surprising that he is entrusted with the education of the children of the neighborhood. Yacouba is the fourth in a family of seven children. His grandfather was a seasonal migrant and was also a merchant in the village, and this was Yacouba's inspiration for migration. Yacouba left his village to go on his first migration in 1966, at age twenty, making the trip on foot to a series of cities and towns in the south. He begged on the streets and studied the Koran in stints of three to four months, in Tessaoua, Gazaoua, Katsina, Kano, Abigee, Sarki, Dan Bouzou, and finally Soura Bildi *quartier* in Maradi, where his children asked around and found the family a house. They have lived here since 1987. Yacouba now has an eye problem that makes farming difficult, but he retains strong ties with the village and visits often, most recently only three months before I interviewed him. On that visit, he took gifts of food and tea. From the village he received a sack of millet, and he always pays the bush taxi fare of those who visit him from home. Like Moussa Diallo and many other migrants, his immediate family lives in Maradi. Yacouba thinks that migration is a necessary adaptation to changing village conditions. Since his childhood, he notes, the village has changed: There are fewer trees, and all the animals except for chickens are gone. Migrating saves food, but also "it has to be a personal decision." He himself prefers the *haské* (city lights).

The nonemployed are sometimes workers down on their luck, but often they are infirm or handicapped. Because it is sanctioned by Islam, begging has no social stigma attached to it in Hausaland and is usually regarded as just another occupation, albeit a low-income one. Only one migrant among the twelve I interviewed in this category stated that he had any cash income at all. This group of nonemployed includes several handicapped people, one beggar, and two retired persons. Mean income is zero, with a median of 41.7 FCFA per day. The average age is 40.3.

Halidou Hamza, fifty, is a blind man from Madata near Madaoua, located 150 kilometers to the west. He left his village for the first time in 1965 at the age of twenty, selling millet for the taxi fare, then coming directly to Maradi. "I have confidence in Maradi," he says. He knew an old migrant in Maradi, a friend from the village, with whom he stayed. Now he lives for free in a straw house in Zaria *quartier*. He does not make very much money, but he is well fed by the community. Caring for the infirm and the handicapped is an obligation required by Islam. Charity in the form of care for the infirm does not differ particularly between city and village, but many handicapped people prefer to be in the city because the greater number of handicapped there provide solidarity. In addition, there are crafts programs for the handicapped in Maradi that are run by missionaries. Halidou maintains contact with the village through visits by members of his extended family. His immediate family occupies a house in the village, and he has fields that he farms with the help of his six children. Halidou had not sent much in the way of food or gifts back to the village in the preceding twelve months, though he did return to Madata once. If he could have stayed in the village, he would have, though he admits a preference for the city.

Occupations as Cross-Classificational Categories

A series of analyses was performed to isolate certain occupational characteristics that give a more detailed picture of the migrants' strategies.

These can then be compared with the overall circular migrant population in Maradi. Older respondents (including guards and community leaders) can take advantage of their long-term residence in the city to gain access to good living and working situations, which may affect the decisions they make about migrating.

Reviewing the years of first migration by occupation reveals that drought can trigger the decision to move. Overall, almost half of those interviewed initially left their villages either during one of the last two droughts or in the year following. The time periods of 1970–1975 and 1980–1985 are particularly prominent for departures, especially for those engaged in the market/sales (43 percent), post-harvest (30 percent), and farming (57 percent) occupations. About 14 percent of the migrants overall were too young to have experienced a serious drought and did not leave the village before 1989.

To some extent, members of all occupational categories seem to have been driven by hunger to make their first migration, though of those involved in post-harvest activities, more than 60 percent left their villages initially because of food shortages; 35.3 percent stated they left because of cash shortage. By comparison, about 36 percent of guards left for monetary reasons. The cash response was highest (42.9 percent) for those engaged in market/sales (though it should be said that many others in this group said they were motivated by hunger). The lowest percentage of responses attributing the move to a need for cash were, not surprisingly, recorded among women and community leaders, who have among the lowest estimated incomes.

In response to the question of how the migrants initially found work, grouping the responses by occupational category reveals that simply "asking around" seems to have worked well for the female occupations and slightly less well for the guards. Relying on family or friends or some other kind of network to find a job was the method used by about half of the respondents overall, and it worked especially well for those involved in post-harvest activities, which revolve around tight groups of people from the same village. Eighty-two percent of those in the post-harvest occupational category relied on friends to find them work. Services and market/sales rely heavily on social networks as well, making this the most common response overall.

Comparing the number of years spent living or working in Maradi for the different occupations gives us a glimpse into the process of migration capital formation discussed earlier. Increased place-specific knowledge affects occupational choices. As Table 6.1 illustrated, the number of years that migrants in various occupations have resided in the city varies greatly, ranging from 7.1 years for post-harvest activities to more than 13 years for community leaders and those engaged in crafts/semiskilled occupations. Subtracting the mean number of years from the year the interview were conducted provides a crude idea as to when the practice of migration became more popular for certain occupational categories compared to others, although it could be argued that this tells us more about individuals' own histories than the role of any outside force. It does support the contention that the prevalence of post-harvest activities in the city has increased considerably in the 1990s, with other categories such as community leaders and crafts/semiskilled displaying a less-precipitous growth. The relationship between time in the city, occupation, and earnings can be better seen in Figure 6.1, which illustrates the effects of duration of residence and occupation on estimated earnings in allowing some migrants to settle into relatively steady, long-term occupations following a volatile period after the initial move.[3]

Whether or not a migrant in the city receives visitors is one mark of how well he or she is incorporated into the social web connecting the city and the village. But whether a migrant receives visits or not is apparently not a strong function of his or her particular occupational groups; there was not a large amount of variation across the occupations. Still, 67 percent of those in the nonemployed category and 76.5 percent of those in the post-harvest activities categories mentioned receiving visitors from the village, as compared with only 21 percent of those in the crafts/semiskilled category.

What season a migrant spends at home is usually a function of his or her access to land in the village and ability to take time off from the urban occupation. Members of all groups, but especially market/sales (85 percent), post-harvest activities (82 percent), crafts/semiskilled (64 percent), and services (61 percent), rely on farming in the home village during the rainy season. Only 28.5 percent of the women reported returning home in the rainy season. Very few of the respondents return home in the cold (December–February) or hot (March–May) seasons.

Taking money and goods home at the start of the rains when they return to plant is another way that migrants channel resources to the villages. Of those engaged in post-harvest activities, the nonemployed, and women, over 50 percent said they sent nothing home. As with the questions about remittances, very little money is taken home when the migrants return to plant. Overall, only 6.7 percent reported taking home agricultural materials like seed and fertilizer. For this sample, at least, this information suggests that at least among this population, migration is not the medium of agricultural intensification that many hoped it would be.

Questions about agriculture were analyzed by occupational category in order to determine whether seasonal migrants contribute to enhancing food security in either their home village or the city. The responses to the

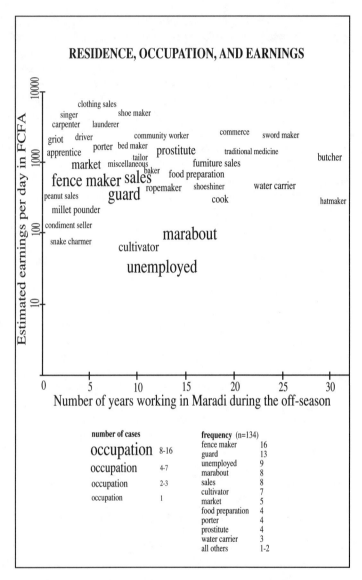

FIGURE 6.1   *Residence, Occupation, and Earnings*

question of whether migrants have fields or not revealed that only guards—who often are Tuareg or Fulani pastoralists and do not normally farm—and women have any tendency toward landlessness. In every other category, nearly everyone still has access to land, and they and their families still grow food in the rainy season.

In examining responses by occupational category about who culti-vates, it should be pointed out that the "self" and "family" categories are not mutually exclusive, that in many cases those respondents answering that "family" farmed the land implied that they too helped with the planting. Bearing this in mind, only 52 percent of the migrants in the market/sales category said they themselves farm, and only 14.3 percent of those in the crafts/semiskilled category reported that they themselves farm. Neither seems as indispensable in the fields, or as in need of farm-ing output, as the post-harvest activities respondents, who reported in 70.6 percent of the cases that they themselves farm.[4]

In response to the questions about what the migrants spend their earn-ings on while living in Maradi, no occupational group stood out as excep-tional remitters. Sixty-seven percent of market/sales respondents, 54 per-cent of service respondents, and 50 percent of crafts/semiskilled respondents reported that they satisfied only their own basic needs with their city earnings. By comparison, only 29 percent of the farmers reported that they paid only for their own needs in the city, and half of them re-ported that they sent money home. The size of remittances is usually quite small, generally in the range of 2,000 to 5,000 FCFA. Over 21 percent of the guards sent home 5,000 FCFA or more during the off-season, compared with 14.3 percent of crafts/semiskilled, 14.3 percent of market/sales, 5.9 percent of post-harvest activities, and 3.8 percent of services. Among the women, the unemployed, the community leaders, and the farmers, large remittances occurred in fewer than 1 percent of the cases. Everyone in-tends to send money home, but except for perhaps the guards, who earn the steadiest income if not the highest, and those engaged in crafts/semi-skilled work, few occupational categories were exceptionally strong in re-mitting money and goods. Remittances by occupation group are presented in Table 6.2, with the cash amounts expressed in FCFA. (Multiple mentions for clothes, food, and "other" categories are included.)

The breakdown of the uses of earnings by income category supports a sentiment expressed by Maradi residents themselves, that the poorer mi-grants tend to remember the suffering in the villages better and to remit at greater rates than the richer migrants, who tend to avoid contact with the village. In doing the interviews, I was surprised that those in the higher income categories (1,500–2,000 FCFA per day) often stated that they could only support themselves and could not afford to send money home. By contrast, many of the migrants in the lowest income category (500 FCFA) stated that they did remit money to their villages, even if the amounts were small. Along similar lines, those in the lowest income cat-egories also receive more visits from the village, and those with lower in-comes tend to take more goods home with them at the start of the rainy season, especially farming materials.

TABLE 6.2 Remittances by Occupation Group: Percent of Individuals in Each Occupation Group Providing Remittances

Occupation Group	Clothes	Food	Other	No Money or Goods	1000 or Less	2000 to 5000	5000 or More
Community leaders (n = 8)	25.0% (2)	25.0% (2)	0.0% (0)	37.5% (3)	0.0% (0)	12.5% (1)	12.5% (1)
Crafts/semi-skilled (n = 14)	7.1% (1)	0.0% (0)	0.0% (0)	42.9% (6)	7.1% (1)	28.6% (4)	14.3% (2)
Farming (n = 7)	0.0% (0)	28.6% (2)	0.0% (0)	42.9% (3)	0.0% (0)	28.6% (2)	0.0% (0)
Guards (n = 14)	0.0% (0)	0.0% (0)	0.0% (0)	28.6% (4)	0.0% (0)	42.9% (6)	21.4% (3)
Market/sales (n = 21)	4.8% (1)	4.8% (1)	0.0% (0)	47.6% (10)	14.3% (3)	9.5% (2)	14.3% (3)
Post-harvest activities (n = 17)	0.0% (0)	0.0% (0)	0.0% (0)	70.6% (12)	5.9% (1)	11.8% (2)	5.9% (1)
Services (n = 26)	7.7% (2)	7.7% (2)	0.0% (0)	30.8% (8)	3.8% (1)	42.3% (11)	3.8% (1)
Unemployed (n = 12)	0.0% (0)	0.0% (0)	8.3% (1)	66.7% (8)	8.3% (1)	16.7% (2)	0.0% (0)
Women's occupations (n = 14)	0.0% (0)	0.0% (0)	0.0% (0)	78.6% (11)	0.0% (0)	14.3% (2)	0.0% (0)

Data on money remitted or taken back to the village at the start of the rainy season support the contention that over the time line of the circular migrant's residence in the city, earnings increase as contact with the village decreases. The data from the post-harvest activities category support this. These respondents have worked in Maradi in the off-season for an average of 7.1 years, and almost half of the post-harvest workers have been coming to Maradi for four years or less. They reported sending nothing home 53 percent of the time, compared with 21 percent of the guards and 15 percent of the services people. The highest percentage of those sending money home came from those in the crafts/semiskilled group, who average 13.6 years in the city. Community leaders stated that they had sent no money at all home during the previous year, preferring to provide support in other ways.

TABLE 6.3   Daily Mean Income (FCFA/day) of Remitters and Nonremitters

Profile Group	Satisfies Needs and Remits Money	Satisfies Basic Needs Only	Overall Mean Earnings
Community leaders	512.5	339	340
Crafts/semi-skilled	1133	1789	1461
Farming	22	425	131
Guards	662	881	751
Market/sales	1810	747	938
Post-harvest activities	365	592	510
Services	1393	779	1087
Nonemployed	500	0	42
Women's occupations	1133	579	698
Overall average	955	709	763

Table 6.3 lists the estimated daily mean income both of those migrants who state that they do remit money home and of those who state that they can only satisfy their own basic needs in the city. The remitting behavior of the migrants seems to vary by occupational group as well as by experience: In some groups, like services and market/sales, the incomes of those remitting money are higher, which would support the notion that these individuals earn more and thus can afford to send money. In other groups, like crafts/semiskilled and post-harvest activities, migrants earning less report sending money home more often. Do the poor remit more in general? If the poor do not remit more, they may be in closer contact with their villages, and this may count for just as much. The relationship between income and remittances is complicated, and unanswered questions remain: First, how do migrants' intentions and behavior vary over the calendar year? Second, how do social networks operate across space (implicating distance and travel time)? Third, what role is played by individuals' occupational histories within the informal sector? And fourth, what is the role played by social contacts and resource flows within Sahelian households in both urban and rural settings?

In assessing the utility of the cross-sectional analysis for occupational categories of Maradi migrants, considerable variation has been observed in the behavior of migrants in different occupational groups, particularly in their relationships with their home villages. The potential of this technique to understand a specific subgroup's behavior in relation to a larger population should be acknowledged. My investigation here points to the heterogeneity in earnings and forms of contact with the rural village, findings that require more analysis with a larger sample and more controlled instruments.

State Interference in Circular
Migration Practices

The story of Garba Garba highlights the precarious existence faced by many seasonal and temporary migrants who live in informal housing in the city and work in the informal economy. It reflects the Nigérien state's resistance toward migration and urbanization.

Garba Garba is the unofficial *mai gari* (or chief) of the Gao migrant community on the southern outskirts of town. He founded the community around 1978, after traveling and working throughout West Africa and in Saudi Arabia. The migrant community of Gao has the reputation within the city of harboring thieves and other criminals. During our initial interview in November 1995, Garba expressed fear that the government would soon force the residents of the community to leave, and he felt that it was perhaps time to move on himself. When I returned to interview him again in May 1996, his wife told Kanta and me that Garba had just been arrested by the Maradi police.

More than a week later, after he had been released, he reported to us what happened on the night of the police raid.

> The police came and looked in the houses at two o'clock in the morning, before *Tabaské* [a major Moslem holiday]. Nine soldiers came into my compound and handcuffed me. They took me out in the street and *stacked* me, handcuffed, on top of several others. Then the army soldiers climbed on top of us. My family had no food at all for two days, only water. The army asked for 4,000 FCFA to release me. I refused to pay anything, so I spent ten days at the police station and received bad treatment while in custody. The police said they were looking for a thief. I think it's because I am a *Bori* [animist] priest. Others in the community were arrested too: my cousin, Lawaly, and the woman across the way.

I asked Garba if he thought the police were harassing them because they are all migrants. "They are bothering us because we live in straw houses, that's why," he replied. He swept his arm across our view of the millet-stalk huts of the migrant *quartier* and said, "Nobody has money, nobody has the legal right to be here."

Are migrants like Garba Garba actually a threat to the state? In some ways, they are because of the challenge they pose to the feeble legitimacy of the state authority. In the view of the mayor's office, migrants pose multiple threats to social order. The Maradi police associate them with criminality and the presence of guns. The Gao settlement is on the southern edge of town, amid numerous paths that run through the millet fields to the porous border with Nigeria to the south. Migrants have also been

Photo 6.3   Balkissa Oumarou rests after her return from the coast. She purchased the radio with some of her migration income.

associated with the spread of disease. In Garba's case, the powers of *Bori* make him an additional threat to an Islam-based social order. Garba is a proud man and decidedly against docility. He rails against the army's idiotic brutality. In the view of the state, circular migrants do not respect borders, they are not easy to locate or count, and they pay no taxes. In Maradi, where a strong indigenous state apparatus has never existed and where people define themselves in opposition to various forms of rule, rural people who freely move about the country are indeed very threatening.

The Resource of Ingenuity

Balkissa Oumarou (Photo 6.3) is a schemer, and one who truly eats the dry season. She is able to do so because she has an alternative source of survival income waiting in reserve. Balkissa has her parents and her extended family back in the Tarka Valley north of Dakoro; she is also married to Boubé Dama, the guard at the Peace Corps hostel in Maradi. Balkissa does not have small children, and this enables her to move much more freely. Like Hawa Hada, another Bororo woman (whom we met in Chapter 1), Balkissa migrates to the West African coast, selling teething medicine for babies and dressing hair. This is her sixth seasonal migration to the coast. She traveled with a group of about twenty-five Bororo women from Maradi to Cotonou, Benin, by way of Dosso in southwestern Niger. In her words, "Getting to Cotonou took about ten days, but once we got there we stayed two months. Lomé was a bust—we only spent a day there and it was difficult—so we left. We took a bus to Accra [Ghana] and spent the next two months in the same compound."

Like her husband, Balkissa grew up about twenty-five kilometers north of Dakoro. She is not formally educated. Her parents were always in motion, taking their large cattle herds north when the rainy season began and back down south in the dry season. Asked if her parents ever traveled the way she does now, to Cotonou and Accra, she replies that they did and that as kids she and her brothers and sisters went along, leaving the animals with other family members up north near the water points. I ask if much has changed in the environment of Dakoro since her childhood. "There is not as much land for animals, and the wells need to be dredged," she says. "During the droughts, even the rich become poor."

Her migrations afford Balkissa and the other Bororo women a chance to escape the harshness of the herding life. "As soon as we arrive in a new city, we all look for a place to stay, where we can leave our things." It should be pointed out that they do not take very many things. I ask Balkissa if she makes money on these migrations. "Well, we eat, and we save a little." The group left around late January and returned in May. Many Bororo women go on seasonal migrations like this every year now. Balkissa feels that the Bororo women get business because they look unusual, and people are attracted to them and buy what they're selling.

I ask Balkissa how she thought Boubé felt about her going away the way she did. She replies that Boubé could see the advantages of the migration as well as she could. Balkissa has an infectious sense of the possibilities that migration can afford. She brings back intangibles, like knowledge and news. She has seen the ocean, and she has found a place where she could enjoy herself and recover her health. Although she did not know anyone in Accra prior to going there, the next time she returns she will have many contacts. She learns a little more each time. Next year she plans to travel to Cameroon, and she says there is now a bus that goes directly from Kano to Douala. Of the money Balkissa earns on her migration, she plans to give about half to Boubé and a small amount to her father. She will save some for emergencies and keep the rest for herself, to spend on clothes, condiments, and food.

Toward the end of our interview, Balkissa states that *she* has a question for *me*. "Do *you* think it is better for people to stay home, or to migrate?" This is a question I had asked her earlier, a question that seemed nonsensical to her at the time. I reply that it is my opinion that migrants do better because they have a spirit in them that impels them to move. Smiling, she is satisfied with my reply and says, "If people migrate, they have the chance of making some money—even if it's just 6,000 or 10,000 FCFA—and they can help somebody."

Balkissa's migratory habits are significant on several levels. As a woman with the freedom to go where she likes and do what she pleases, she is unusual in the region. Balkissa has turned her own ingenuity into a valuable resource. In the context of the structures that were developed to describe changes in Maradi Department, Balkissa's experience resonates deeply. Though not formally educated, Balkissa is what is called in Hausa a *matan zamani*, or "modern woman." In demographic terms, she has fewer children than the regional norm, and she is likely to have exercised more choice over her reproduction than many other women. In political-economic terms, although Balkissa as a female migrant is invisible to development efforts, she is critical to them because she has both escaped the bonds of the rural household and helped stretch norms. In environmental terms, she has employed flexibility and knowledge to adapt to changes both in the grazing lands her family occupies and in the network of West African cities. In social terms, Balkissa has skirted oppression by moving out of the vise that rural life can be, and she has used mobility to gain access to the world and to satisfy her needs.

For demographic, political-economic, environmental, and social reasons, Balkissa is a model for future Nigérien women. As will be seen in Chapter 7, women have shouldered much of the burden of the changes occurring in the rural sector, where population growth compounded by environmental change and political-economic change have presented a profound challenge for the future of the region. Women desperately need a model, and they need to move into spaces where they can make decisions governing their lives. Change of this sort, however, which challenges the life ways of a people, will have its consequences.

Traditions

I gathered a group in Maradaoua *quartier* in December to discuss some long-standing questions and issues that had arisen during the individual interviews. Maradaoua is an established part of Maradi, and it has fewer temporary migrants living in it than do *quartiers* like Gao or Zaria. Kanta had set up a meeting in advance, and after we arrived, we gathered some men together, including the *marabout* whom I had already interviewed, Malam Harouna. I had some remaining questions that I wanted to ask. I have already mentioned that even with Kanta sitting next to me, the weight of the group against me was making me feel uncomfortable.

I began by talking about the indeterminate identities of some of the migrants. Are the *masu cin rani* (seasonally displaced) and *masu digga* (temporarily displaced) strictly the poor of the village, or do rich people also make short-duration migrations? The Hausa do not have elaborate categories for the different social classes; one is generally either *sarauta* (royalty) or *talakawa* (the poor). For the most part, those in the group responded, migrants are *talakawa*, with only a few being comfortably well-off. One man replied that "almost everyone doing seasonal migration is hungry" and that "a man will sell his family's own food to pay for the bush taxi ride to the city." These confirmed my own impressions. But then, I asked, should seasonal migrants be considered "enterprising" people, who are moving in response to economic opportunities in the city? They disagreed with the enterprising characterization, seeming to say that the root cause was less important than the pressing need to leave the village to find cash or food.

Who maintains better contact with their villages of origin, the less well-off or the better-off migrants? I told the group about my impressions of the migrants, and I related that most migrants seemed to be in close contact with their villages, clarifying contact to mean getting or supplying news to the village and satisfying requests for support or mutual aid from the village. Kanta had suggested to me earlier that the better-off, more permanent residents of Maradi fear contact with the village because they are wary of the demands placed upon them by loose acquaintances, dependents, or other hangers-on. On this topic the group was unequivocal: The poor maintain much better contact with the village than the rich. The rich try to cut their contact whenever possible. Everyone agreed vigorously that the *masu kudi* (those with money) mostly help their friends in the city, instead of their families or their acquaintances in the village. They might consciously restrict contact with the village in order to avoid getting pulled down by it. This does not go unnoticed by those who continue to live in the village.

What about the importance of Maradi-*ville* as a destination within the surrounding region? If, hypothetically, the city somehow ceased to exist, what would happen to the villages around Maradi? In the past I had discovered that often such conditional questions were mistranslated and misunderstood, but this time the group seemed to grasp my intentions immediately. They replied that if Maradi suddenly disappeared, migrants would simply go to Nigeria instead, and they would probably not come back. There is a difference, the group reported, between migrating to Maradi and migrating to Nigeria. If migrants come to Maradi, the migrants eat, but they do not take money home. Nigeria is different: There is a real possibility of making money there. There is abundance. Malam Harouna recited some of the features of Niger's huge neighbor to the south: "There are yams sitting in the ground, maize, sauce, salad." To myself, I recalled my vacation in Kano when I was a Peace Corps volunteer, staying in an air-conditioned hotel and washing down my breakfast of steak and kidneys with a bottle of Guinness stout. I asked them if they thought Maradi in fifty years would be part of Nigeria. Never, they replied. Why would Nigeria even want Maradi? "There is nothing here," Malam Harouna said. "No water, no food, no trees, nothing!" I asked the men what they thought would happen to the village of Chadakori if everyone left. Would the village still survive? By "village" I meant the physical inventory of structures, the well and the fields. But to the group, "the village" was the people, and nothing more. If everyone in the village leaves, then of course the village leaves as well. I realized that beneath the semantic differences, and the ever-present problems of translation, I had touched upon an essential kernel of difference: *Places* here are different.

I related some more impressions from my survey-interviews, in which I asked questions about remittances to the migrants' villages of origin. I remarked that even though some migrants earn enough to buy fertilizer and perhaps even an oxcart, few consider such agricultural goods to be wise investments, and few seem to perceive that the millet yields could be improved with additional inputs of time and plant nutrients. They obviously need the food. Why, I asked, do people not invest money in the improvement of agriculture in the village? The group's answers were not enlightening. "There's too much risk of drought," they said. "You just can't do anything," someone said. The problem of finance also figured in their explanations: "There just isn't any money in the villages." I would explore this question more fully on my next trip to Guidan Wari.

Next, I sought to probe whether the group felt that Niger's state of crisis was caused by the current rapid demographic growth. I asked whether the group saw problems with a population that is growing too fast in the village, problems such as crowding, land shortages, soil exhaustion, and food shortages. To my surprise, this question struck them as both bizarre and obvious. Impatiently they replied, "If there are a lot of people in a village, *then conditions must be good*, or the people would leave." That is, the fact that there were a lot of people in a village was the proof of the village's success, an endorsement of favorable conditions. A growing village is not in crisis.

I persisted in my questioning and tried a different approach. Here's a village, I said, drawing a circle in the dirt at our feet. There are people, land, and food. A certain amount of food can be grown on the land. I could almost feel the ghost of Malthus in our midst. If there is the *same amount of food* but *more people* wanting to eat it, won't some people be hungry? No, they replied, "because for every baby that Allah gives us, He takes someone away." Takes them away, how? I asked. "They die." But the number of babies born is not equal to the number of people who die anymore. Don't people also leave when the conditions in the village are good, I asked, but when there are too many people and there isn't enough food to eat? Malam Harouna shook his head. The *anasara* just doesn't get it.[5] When people leave, he said, it has nothing to do with conditions. Malam Harouna looked at me and saw something that enraged him. He began to preach. He derided what he called *"planning,"* by which he meant family planning, as a crazy way to get black people to stop having children. I noted that of course the Hausa language has no word for the concept of family planning and has to borrow English. He shook his finger at me. "You! You know nothing about survival! Nothing!" I decided to move on to the next question.

What happened, I asked, when the drought occurred twelve years ago? Do you agree that many people came from the countryside into Maradi-*ville*? Yes, they replied, a lot of people came into Maradi. Many stayed with friends and families in the city. The city was very full. A lot of people were sick, and some died. Malam Harouna knew one man in Maradaoua who was so hungry that he ate his *roof*. When the next drought comes, I asked, will a lot of people come into Maradi again? The group's response surprised me. It grew very quiet, and I realized I had made a big mistake. In fact I had done something extremely rude (to assume there will certainly be droughts in the future is taboo). After a minute of tense silence, Malam Harouna spoke up. "There won't be another drought," he said with an air of finality. But what is the cause of drought, I asked, feeling that my nerve had better not let me down. "The cause of drought is a lack of rain," Malam Harouna replied. Why doesn't it rain here, I asked. He replied, "Because people don't pray enough. Nobody respects family anymore. There are too many *mutum zamani* [modern people] around, and they are the problem."

At this response I felt a shock of recognition. I had heard and read about attacks in which *marabout*s publicly humiliated women wearing short skirts and blamed their behavior for the recent droughts.[6] This was also the contention of fundamentalist sects in many southern Nigérien and northern Nigerian cities and of the new *Maitatsine* millennial movement, with its prognostications of imminent doom. Such sects appeal strongly to the audience of young urban migrants, who are uprooted from their families and villages and seek wisdom in the words of old men. Though denouncing modernity, the prophets use its technology to spread the word about the "real causes" of popular discontent and to point fingers at the West. I recalled that the revolution in Iran had used audio cassettes to smuggle in the voice of the exiled Ayatollah Khomeini.

In personal terms, I came away with far more from my encounter with the Maradaoua group, though, than an outraged sermon from a *marabout*. I placed my meeting with Malam Harouna and his friends in the context of the "rapid rural appraisal," which normally includes such encounters between a development worker and "typical villagers" to elicit their views. These meetings, which are sometimes called "focus groups," are used to collect social or environmental histories, attitudes about various beliefs or practices, and public opinion; I have made some use of them myself. In the context of development, such encounters have always seemed to me to be rather staged, done for the benefit of the development worker.

I have often wondered how focus groups seem from the perspective of those who have been gathered to express their opinions. I fully expected the villagers to be able to read the real purpose of such meetings, which is often to ascertain the suitability of a particular village for a well or a pharmacy or some other project. These projects are usually seen by the villagers as *cadeaux* (gifts) from "us," the more fortunate, to "them," the less fortunate. With their eyes on the goal of securing the *cadeaux*, villagers must debase themselves and tell their sad stories: "Oh yes, the desertification here has been terrible!" "We're awfully poor!" "We would really like your aid." They understand that for "us" to give "them" the *cadeaux* ultimately makes "us" feel better. This is precisely what Jean Copans meant when he wrote that so-called humanitarian aid is actually an effective means of preserving the dominance of the West and of promoting a beggar's mentality among those who "deserve" our largesse. Aid turns people into victims. It also distorts, and sometimes even erases, the actual process by which "we" became the gift-givers and "they" became the gift-receivers.

This is why I came away enlightened and elated from my meeting with the group of Maradaoua residents. They had the temerity to take the format of the group encounter and turn it on me. In the group discussion, I had the memorable experience–which I had not had since my Peace Corps days—of representing the West and speaking for the oppressors. I also had the experience of being made an object of ridicule when I cast doubt on their knowledge of their surroundings and their ability to adapt. They rejected the slogans of development, and they were fatalistic in their rejection. The men did not care that they were violating the official development protocol.

Instead, in their words and expressed beliefs, I heard voices from the deep past, from the culture of continuity and stoicism. Culture must endure when life is hard. It simply must go on. People die, and smart and tough people survive, becoming the ones who have the families and reproduce the culture. What is "culture" anyway, but reproduction: of material conditions, of the practices of growing and eating food, of passing time with one's family and friends, of making love, of raising children, of caring for one's elderly parents, and of moving when conditions are bad? This is the culture of the village hearth, which provides the comforts of knowing that everyone and everything has a place.

What *will* people do when the next big drought comes? Some will stay and endure it, some will die, and many will be hungry. Embroidered into the tapestry of the culture is the distasteful but absolutely necessary need to cull the weak from the population. The responsibility for this is too fearsome to be accredited to any mortal, so it is given to Allah, but the circumstances and actions belong to the mortal ones, the survivors.

Clearly this age-old culture knows how to survive, yet this is not the problem. The role of the outsider should be to press beyond questions of simple survival, and the role of development should be to provide more than simply Malthusian options. Drought prevention is a matter of human rights, especially for those with smaller voices. In encountering this group of men, I was tempted to throw my hands up and say simply that they know better because they have lived here longer, but this neither addresses their fatalism, be it culturally appropriate or not, nor allows for their ability to change.

Notes

1. During my stay at the Maradi Peace Corps hostel, someone managed to climb over a twelve-foot wall, enter the building, and steal the box containing the volunteers' rent and utility payments, which prompted the proprietor to affix fresh shards of glass to the top of the wall.

2. Occupational choices for women are limited. Women can either make and sell food or work as prostitutes. The latter is five to ten times as profitable, though hardly without its obvious risks.

3. With a larger sample size, a regression analysis could be performed on the effect of occupation and duration of residence on earnings while holding age and sex constant. Such an analysis could be the focus of future work. A second item of future research could examine the transitions in individual migrants' occupations as they become acquainted with urban social networks and gain migration capital and social capital.

4. The variation in migrants' direct participation in agricultural activities supports an opportunity cost explanation: Because they earn more than their family members, some migrants elect to occupy themselves elsewhere and let someone else undertake activities such as preparing the fields.

5. *Anasara* is a Hausa word used in derision for those of European extraction.

6. This news item, as reported in the Niger Internet group, is of interest:

"Public Prayers for Rain

"On Sunday, July 6, 1997, several hundred people, including General Mainassara, prayed for rain at Niamey's Great Mosque. Those in attendance wore their oldest clothes, because, as one *marabout* explained to AFP [Agence France Presse], 'to beg for God's mercy, one must avoid luxury and become very humble.' The prayers, led by the mosque's Imam, El Haji Ismael, lasted for about half an hour. On the previous Thursday, the general made an appeal to Islamic organizations, which organized special prayers throughout the country. At several Niamey mosques, all night vigils were held from Thursday through Saturday, with some people fasting and giving alms as a sign of devotion. In Niamey, there had been no rain for two weeks before last weekend, but, following the special prayers, the skies over Niamey were filled with large black clouds.

"People offered various religious interpretations of the drought to the press. The leader of one Islamic association, interviewed on Tele Sahel, declared, 'We have committed too many sins, God has decided to punish us.' Another religious leader claimed that 'adultery, theft and intolerance are the cause of this drought.' Still others expressed the belief that politics was involved, saying 'God is punishing us because the politicians on all sides refuse to unite in order to save Niger.'

"In Zinder, the previous Friday, religious youths roughed up some girls dressed in mini-skirts and made death threats against prostitutes, who, they claimed, are responsible for the decline in morals"(Agence France Presse, as reported by *Camel Express Telematique*, July 12, 1997).

7

Rural Community
Perspectives

Watching bush taxis fill at the Maradi *tasha* in late May, the morning after the first substantial rains fell, gave a flavor of the changing relationship between the city and rural villages. Located next to the market grounds, the *tasha* is one of the control centers of the Sahelian economy. Huge ziggurats of sacked millet rose from behind the puddles, merchants' shacks, and parked buses and trucks (Photo 7.1). Speculation at this critical point in the agricultural calendar is driven by inflated grain prices—since high-stakes gambling on the climate is an unofficial national sport in Niger—and sacks of fertilizer were stockpiled for the upcoming rainy season. But for the seasonal workers of Maradi, time was on hold. They were making arrangements to return to their villages to plant, but many were not leaving yet. Men crowded around one battered radio to hear the announcer read the rainfall totals for each *arrondissement*. In most areas, not enough rain had fallen yet to start planting. In fact it would be another week before the migrants headed back north. While they were up in their villages, the city of Maradi would experience droughtlike conditions, for initial rains were delayed for many tense weeks.

Located on the periphery of Maradi Department in the center of the Sahelian zone, Guidan Wari and surrounding villages have seen some rather dramatic changes in the conditions of agricultural production and in access to land and other natural resources. The Sahelian zone as a whole has experienced profound transformations in its internal attributes. But this alone does not explain how Guidan Wari changed during the 1980s and 1990s. Increasingly, its future is being determined by people's access to markets and opportunities, particularly to urban opportunities. Guidan Wari has come to resemble the villages of Tahoua Department in central Niger, where food self-sufficiency has been achieved

Photo 7.1   Grain piles and traders at the Maradi tasha *show that gambling on the rains is a significant component of the national economy.*

through influxes of income from migrants. Like Tahoua, the village anchors a region characterized more by economic losses than by gains.

In the course of my interviews with migrants in Maradi-*ville*, I became acquainted with some who came from Guidan Wari. During my stay, I made a series of trips there—including a memorable one on a mountain bike in the dry season—which acquainted me with the people and the terrain of the area from which I took much of the information for this chapter on Guidan Wari. This chapter is therefore less an exhaustive study than a series of impressions.

Circular Migration in Context

The desultory nature of rural agricultural activities at present and the nagging sense of obligation that some part-time urban residents feel toward their home villages underscores the marginality of their places of origin. The small satellite villages surrounding Guidan Wari are located in the "periphery within a periphery" of Maradi Department, but in fact they suffer from triple or quadruple relational disadvantages, disadvantages that can be observed at multiple levels of analysis. They are satellite villages in orbit around a larger village, located in a dry district in a sparsely populated arid country, within a large and economically diverse macroregion, on a poor and marginalized continent within the world system. As a peripheral region within the migration system of Maradi Department, Mayahi *arrondissement* is typical in its use of the fluid medium of labor, which has had the effect of pulling the settlements toward greater interactions with the outside world.

The spatial pattern of the migrants' sending villages reveals the dual effects of geographical *site*, which are village conditions relating to agricultural potential, climate, water table, and soil, and *situation*, conditions that relate to proximity and market access. Migrants are created by both site- and situation-based conditions: In the first, migration streams are driven by food insecurity or labor or land crunches in the village, and in the second, migration streams are driven by proximity to markets and opportunities.

It is not accidental that well over half of the circular migrants interviewed in the city of Maradi come from the most recently transformed Sahelian districts, particularly from Dakoro and Mayahi *arrondissements*. These districts were the last settled within the department, are the farthest away from the infrastructure and opportunities of the city, and generally have the poorest access to resources. Within these districts, many of the migrants' villages of origin are in fact small hamlets—sometimes, in fact, too small and scattered to have been included on the 1988 census. The prevalence of circular mobility varies inversely to agro-ecological potential here because mobility is the least risky of economic strategies and is well suited for an area of such unpredictable conditions. Migrants from these northern districts are often temporarily displaced men who have left their families in the village and who work in sales or post-harvest activities in the city.

The remaining circular migrants are not from the distressed districts of the north but, instead, originate in the southern villages in the vicinity of the city of Maradi. These villages lie along the axis of east-west movements and maintain contact with the Maradi hub through a network of bush taxi and truck traffic. The prevalence of circular mobility here varies by the level of access to urban markets and opportunities. Migrants from the southern districts are often either seasonally displaced or more-permanent residents of Maradi who have moved their families to the city and who work in crafts, services, or sales in the urban economy. It should not be forgotten that additional factors—such as the subregional ordering of village settlement, availability of land, access to markets, the degree of social cohesion, and the power of the chief—all play roles in determining mobility prevalence.

The changing rural-urban relationship in the Sahelian zone has been recounted as a geographical *process*. For Guidan Wari and other villages like it, the developmental process of regional change—settlement and population dynamics, land-use intensity change, and political-economic change—is dynamic, circular, and cumulative over time, and it advances in two parallel streams: the site-based erosion of rural self-sufficiency, based on the in situ relationship between population and environment; and the situation-based reach of the city, which engages the village in new relations. Both of these streams are partially implemented by mobility. In the historical geography of settlement of the Sahelian zone, mobility has occurred as a medium of change, in phases, over a time line. These phases can be described in the following sequence.

The first phase was the *frontier phase*, which occurred in the Sahelian zone from the early nineteenth century to the 1920s and 1930s. As was recounted earlier, the land was initially uninhabited or thinly populated; settlement was restricted for reasons of conflict and climate. In his seminal book about the origins of African societies, Igor Kopytoff characterizes the internal, or local, frontier as a natural force for cultural transformation, and also as "a political fact, a matter of definition of geographical space" (Kopytoff 1987, p. 11). Hausa and Bougajé peasants settled on the Sahelian fringe, and one day at a time, season after season, they re-created their culture there. Colonial rule had the effect of furthering the process of settlement, by removing the threat of raids and allowing people to spill out of their guarded refuges and spread across a landscape dominated by large game and plant-covered sand dunes. Population densities during this phase were low, and mobility types included the caravan trade— which the French valiantly tried to obliterate—herding, and hunting, as well as the movements associated with shifting agriculture.

The second phase was the *bush consumption phase*, which occurred from the 1920s and 1930s to the 1960s. Driven by natural increases in the population, settlements continued to spread across the Sahelian zone, and bush land was gradually cleared and converted into agricultural fields. This created an increasingly inhabited landscape, which was marked by the spread of markets and roads at the expense of game animals and bush foods. The colonial enterprise encouraged new forms of exchange, and the groundnut cultivation that the French promoted was enthusiastically embraced by peasants. Gradually, the bush land buffer that was critical to the mitigation of rainfall variation was consumed. The state of systemic vulnerability worsened. The analysis of land-use intensity changes documented the gradual filling of Mayahi *arrondissement*, beginning first in the wetter southern agro-ecological zones and then spreading into the drier zones that are more marginal for crop agriculture. When village conditions were compounded and aggravated by drought, hungry peasants left their villages and sought family and friends in the city, or they begged on the streets. As the crisis continued, mobility triggered by drought achieved unprecedented levels.

The third phase was the *circulation phase*, which began in the late 1960s and continues in the present. Once movement had been triggered by drought, peasants gradually became habituated into urban social networks and gained a foothold in the urban economy of nearby cities. Yet every year in the rainy season, they continued to return to their villages to plant. As more and more rural people circulated seasonally and participated in the cash-based economy, a nonlabor counterstream consisting of goods, cash, food, tools, other materials, and knowledge followed them back to the rural sector. The analyses of the migrants in Maradi examined their social networks, and it was found that circular migrants maintain enduring ties to their home villages by receiving visits from friends and family and by sending home gifts. But however involved they are in village welfare, the majority of circular migrants are not investing in village agricultural improvements. Instead, they are (re-)building social ties, honoring family obligations, and exchanging news and information.

The urban influence on the rural sector is sometimes confused over the subject of "investment," which normally includes flows directly related to agricultural production but does not include the maintenance of social networks. The French colonial officers perceived the Hausa economic philosophy, a philosophy based not on accumulation but on the permanent redistribution of wealth (Fuglestad 1983). Groundnut profits were not invested in what the Europeans regarded as productive undertakings; rather, they were spent on gift giving and other activities to maintain the social network. The French overlooked the rationality of this philosophy, but it should not be overlooked by those concerned about famine vulnerability. People, and especially powerful people living in areas not afflicted by drought, are most useful in times of societal crisis. This is especially true when outlets for financial investment are limited and suspicions about the state are strong.

As I have established, the contexts for mobility follow individual developmental paths and are contingent on other contexts. Equating these underlying forces *directly* with migration motivations is obviously facile and misleading. No one decides to migrate because of "land-use intensity" change, and few leave their families behind because of "population growth." Instead, these underlying forces should be seen as deep explanations, what some would call discourses. The cognitive gap between underlying forces and individual responses grows as one scales down to the level of the village. As one approaches the human scale, the contexts converge into a lived unity that is based subjectively on individual interpretation.

In Guidan Wari and in some of the smaller villages surrounding it, I asked residents to estimate the proportions of the various kinds of circular migration, as well as their destinations.[1] Some mobility types (such as the market or seasonal agricultural wage labor mobility types) for Guidan Wari, which I discussed earlier, did not figure in the villagers' estimates. These will be covered separately later in the chapter. The mobility prevalence data for Guidan Wari *terroir* are presented graphically in Figure 7.1. As the figure reveals, the villages within the *terroir* with the lowest overall levels of mobility for men are the villages settled earlier, such as Guidan Wari and Guidan Wani. The villages with higher mobility prevalence were settled later and have fewer nonagricultural opportunities, poorer access to natural resources, and more environmental stresses in general.

Informants stated that in the villages surrounding Guidan Wari rates of circular migration have been high since the *Dan Koussou* (1968–1974) drought. Since the villagers were forced to sell off their animals, there has been nothing left for them to do in the dry season. Crop yields have been low, and there are no work opportunities in the village. Families never fully recovered from the drought, so more and more migrate every year. High levels of mobility have persisted since the drought.

As the migrants in the urban realm of Maradi also stated, the choice of destination is governed by proximity and cost considerations. Many like to go to Maradi because they are still nearby and can get news from their village. It is simply more risky to go farther away, though men from some of the villages migrate to Nigeria, Cameroon, Libya, and Arlit (the uranium mining town in northern Niger). The men who left the villages and got married in Maradi now return only for celebrations. When they come back, they bring money and clothes, but they rarely bring chemical fertilizer home because it is too bulky and expensive. The young Koranic scholars go to Maradi or Tessaoua to study and to beg. If they are in these nearby cities, they come back for weddings and naming ceremonies. Informants also affirm that ties to family are easy to maintain remotely through contacts and visits. Migration money buys millet and animals, which is very helpful since there isn't any money to be made in the village.

For the villagers, the connection between local conditions such as bush land loss and the high rate of circular migration is clear. The villagers maintain that if there were good land, they would migrate less, but they explain that having no money is more important than having no land. If more men stayed to work in their fields, they would have more food. The villagers agreed that it would be better to stay and work in their fields, but said they are compelled to leave because there is no money. The villagers used to raise and sell animals to buy food when necessary, but the loss of their herds has eliminated this safety net. Now they are desperate. They cannot save anything, so any cash they earn is applied directly to-

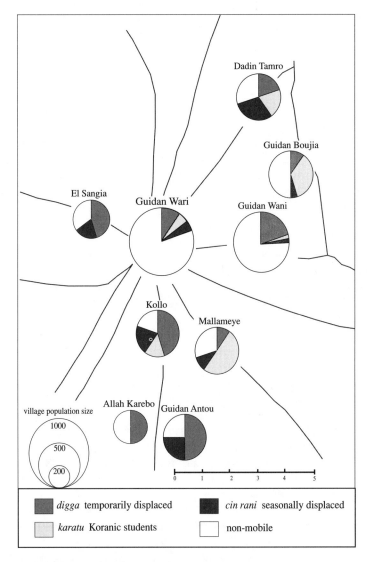

FIGURE 7.1   Circular Mobility Prevalence

ward food purchases. In these villages, there was a stated connection between the lack of economic opportunity, the absence of free land, and the worsening crop yields.

Among the consequences of rapid urbanization along the Sahelian zone is that places like Guidan Wari are increasingly left behind because food and money-making opportunities are lacking. In the rush toward the city, the village is left with a skeleton crew, and few opportunities exist other than continuing to work the land. Four detailed responses to increased levels of circular migration are described next.

The Land-Use Change Response

It is clear to the villagers in Guidan Wari *terroir* that the local environment cannot tolerate the present intensity of use. The problem seems obvious to the villagers: There is nowhere else for cropland or settlements to grow. When I asked the villagers for solutions to the soil mining that has occurred from continuous cropping, they responded that the lost nutrients had to be replaced. The soil in the region is sandy and poor in organic matter. Manure from livestock herds helps replenish the fertility. Some villagers in the region pay the Fulani herders to lead their cattle over the fields to fertilize them, but often this alone is not sufficient. Nor is the aeolian deposition of dust from the Harmattan wind in the winter months sufficient to raise soil fertility levels.

The young chief Boukary Illa maintained that in 1995 there had been enough rain for the crops but that the yields were still too low. He also maintained that the village would be self-sufficient in food if the farmers could just maintain the soil fertility. Boukary jokingly said that the field across the road where people go to defecate has some healthy millet plants. Chemical fertilizer (*takin zamani*) is experiencing increased demand here. However, the villagers balked at the astronomical price of a fifty-kilogram sack of fertilizer (which costs 4,000–5,000 FCFA, or US$8–10). Farmers have also begun to collect more animal manure for their fields. Carts and oxen for distributing manure are popular investments in Guidan Wari, and of course the cattle themselves are a fertilizing agent. Boukary estimated that there are ten oxcarts in the whole *terroir*, which contains about 3,500 people, with all but three carts in the village of Guidan Wari itself.

Livestock, including cattle and, more commonly, sheep and goats, remains the most common form of investment in the village. Livestock is also the most economical means of replenishing the soil nutrients mined from the soil through continuous annual cropping. Before the 1968–1974 drought, the villages had far more animals. Most of these animals survived the 1968–1974 drought but not the 1982–1984 drought. Some families have succeeded in building back their herds to predrought levels, but many other herds have never recovered.

Between 1994 and 1996 the prices for chemical fertilizer in Maradi doubled, from 2,500 FCFA (US$5) in 1994 to 4,500 FCFA (US$9) in 1995 to 5,500–8,000 FCFA (US$11–16) when I was there in the spring of 1996.[2] One informant commented that he thought the price was high because the fertilizer was being smuggled out of Nigeria, where it is produced, and routed through Niger to Mali where it is used for rice cultivation. Exporting fertilizer out of Nigeria is illegal because demand there has outstripped the government supply-and-regulation efforts. Chemical fertilizer in Maradi Department is most often used for groundnuts, for which it is mixed into the soil during plowing; fertilizer is used less often for millet.

Guidan Wari is a land-use contradiction. Chemical or organic inputs are necessary now to maintain the fertility of the soil. Villagers can no longer simply put their seeds in the ground and expect them to grow. Given existing technologies, population density needs to be low because yields are low and people have to be able to walk to their fields. Yet crops require inputs, and it does not appear that the need can be satisfied with on-site manure production. Increasingly there are too many people to be fed using the traditional methods of shifting cultivation, but there are not enough people with sufficient incomes to support a market for what have become agricultural essentials. This fertilizer gap continues to lie at the root of the region's food sufficiency problems.

As Chief Boukary said, even in good rainfall years now, the yields are insufficient to feed the population. The new importance of road connectivity cannot be underestimated, and significant changes have occurred since the construction of the new laterite road connecting Guidan Wari with the cities of Mayahi and Maradi in 1992. The potential for a local market is present but largely unrealized, particularly for fertilizer and oxcarts. Were the market in Guidan Wari to be somehow freed from dominating economic constraints—such as the lack of investment capital, the relatively small population in the area, the distance to markets, the perishability of goods, and the risk of drought—the demand for fodder, oxcarts, manure, and chemical fertilizer would be astounding. A potential market already exists for fertilizer, oxcarts, and other tools of intensification, but when I asked a Maradi merchant why he or someone else did not ship a truckload of chemical fertilizer up to Guidan Wari and sell it, he snorted and replied, "What's the use of sending it up there? Nobody has any money!" Such an attitude, driven less by prejudice than by knowledge and experience, seems fated to condemn Guidan Wari to continued agricultural stagnation.

The Livelihood Response

Livelihoods are themselves a response to uncertainty and variability. What began as rural occupations were pushed out to the city because of drought, which forced villagers to pick up and move. When able-bodied and enterprising members leave, the village is deprived of their energy and resources. It should not be forgotten that when households in the Sahel become integrated into the market economy, their ability to withstand food shortages precipitated by drought is not automatically eroded; often it is enhanced (Curry 1989).[3] This diversification is a response to the likelihood of drought, and it spans the Sahelian zone. If the practices adopted by villagers in Guidan Wari are allowed to speak for them, the strategy of satisfying food needs by migrating seasonally has become the method by which villagers avoid starvation and improve future abilities to respond to uncertainty and risk in their vicinity.

In Guidan Wari, many men supplement their incomes with other work during the cold and hot seasons. In the village there are tailors, merchants, grain mill operators, a chief of the local taxi-park, carpenters, bakers, butchers, launderers, water vendors, medicine vendors, and blacksmiths (Jones 1995). Though these trades did not necessarily evolve through contact with the city, the parallels with urban occupations should not be missed.

Aside from their responsibilities in the household and the fields, many women in Guidan Wari also work to make supplemental money. There are a number of women who sell water as well as foods such as bean cakes, millet cakes, beans, pounded millet, millet broth, and spiced millet and milk broth. These foods are sold by daughters or older women at the market or during off-market days. There is also a female tailor in the village. Older women who do not have young children may have more time to engage in such activities. Otherwise, in order to earn cash, women have to find time when they are not needed in the fields or by their families. Again, as with the male occupations, these livelihoods parallel urban occupations because market relations and the need for cash have spread to the village.

The solution to the downward-spiraling local conditions of production as described here appears to lie in developing *more* opportunities to raise cash in the nonagricultural sector to facilitate *more* purchases of food and essential goods, not fewer. Ultimately, the virtue of circular migration is its potential ability to return resources directly to the village in a timely manner. To comprehend the utility of circular migration behavior at this level, we must ask not why people initially left their villages but why, after migrating, they continued to maintain close contact with the villages.

As we have documented through the changing calendar in the Sahelian zone, time and labor are valued differently now in villages like Guidan Wari. Through growing market involvement, villagers have been pulled away from customary obligations and toward a season-specific accounting of labor allocation. The organization of key activities such as selling grain after the harvest, speculating on prices prior to the rainy season, periodically replanting and weeding, and the mobilizing labor for clearing fields, burning bush, planting, and cultivating are all consequences of the shift to increasingly careful forms of time management in an agricultural calendar that is competing with urban livelihoods. The shift is compounded by the lack of rootedness of rural occupations, which has facilitated their transplanting to the city.

The Social Network Response

One implication of high levels of circular migration is that more and more people from any given village are not in fact living there anymore but are maintaining contact from a distance. We have seen evidence of the extensive use of social networks to help migrants find food and lodging upon arrival in the city and to find work after settling in. Many occupations operate on the basis of personal contacts and word-of-mouth as well. Nearly all migrants I interviewed were in contact with their villages, hosting visiting friends and family members, sending cash and other gifts home, and themselves visiting their village homes.

This level of interaction indicates a deep level of involvement by urban residents in the daily economic affairs of the rural village, which has important implications for drought preparedness. The most surprising finding from Hopkins's and Reardon's (1993) research on Sahelian rural household budgets was that total rural household incomes did not vary by agro-climatic potential; rather, the proportion of total household income supplied from nonagricultural sources, including migration remittances, varied. This finding implies that in the villages studied, migration remittances serve the needs perceived by the migrant (as opposed to those in the village) for the maintenance of a subsistence minimum in the village.

The result of the village-regulatory function performed by absent village members is a "concerned absentee" phenomenon, made possible by large numbers of migrants who are no longer present in the village on a full-time basis but who monitor the village from afar. A social web connecting the village and the city enables information about the well-being of family and friends to be transmitted. During times of crisis, news about the village reaches outside to absent members in the cities, and urban residents respond. Money or food is transferred home, and the village recovers. In a sense, this social web connecting the village and the city enables a kind of remote control to operate, where *just enough* resources are provided to keep a village functioning through droughts or seasonal shortfalls. Through the mechanisms of circular migration and social networks, food security in the rural sector can thus be maintained. Clearly the level of diversification of income sources in the city and the higher levels of earnings overall play a large role in fashioning the nature of responses.

In Guidan Wari, many kinds of contact with the outside are maintained, some of which can be roughly estimated. Both circular and more permanent forms of mobility play a significant role in buttressing the village's arsenal of drought-mitigation strategies. Outside connections, especially those to wealthy and powerful relatives, have the value of providing a kind of safety net for village members. One of the sons of the elderly chief of Guidan Wari, for instance, is a well-known doctor in Zinder. For years, parents send their children to school, and absorb the loss of their labor, in the hope that their formally educated offspring will succeed in the civil service system and support the parents through crisis times.

Figure 7.2 maps two common migration types by residence within Guidan Wari village, illustrating the spatial pattern of families with social network ties outside the village.[4] The two types of mobility depicted are temporary mobility, which includes seasonal and temporary displacements, and what can be considered more permanent mobility, which includes Koranic study. A spatial pattern can be detected in the prevalence of each of the two types, which corresponds significantly to the process of village formation. Villages normally grow from the inside out. More established members arrived earlier and located themselves toward what becomes the center of the village, whereas later-arriving members found accommodations along the edges of the village. Not coincidentally, families with "sons" who participate in the mobility types that are considered more temporary are located around the edges of the village, whereas those whose "sons" participate in the more permanent types of mobility tend to be in the center. A similar spatial pattern can be seen in the residential preferences of circular migrants in Maradi-*ville*. Among the observations that can be made about this spatial pattern is the apparent class differentiation in the role of Koranic study in introducing young men to the social channels of the city.

Through the process of urbanization, mobility serves to fill the gaps in production caused by seasonal and annual shortfalls. The process through which this occurs is the continuous monitoring of the village by absent members. This surveillance keeps the village viable, but just barely. Nonresidents keep watch on the village and inject cash or food when a crisis deepens, always providing just enough to keep operations running but never enough to pull the village out of its subsistence crisis. Such affective responses in the Sahelian zone characterize contacts that are maintained less out of economic interest than out of concern or obligation. As a result, the development of the rural sector is intentionally arrested. For real gains to be made in the villages, the entire rural-urban relationship would have to be reconfigured. This is another way that

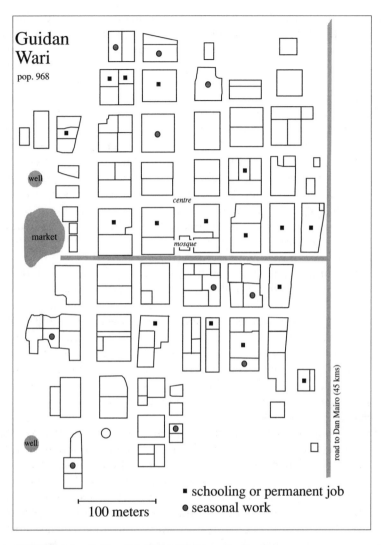

FIGURE 7.2   Guidan Wari Mobility Types by Residence

mobility reduces rural famine vulnerability in the short term, but with unknown long-term consequences for individuals and communities.

The Demographic Response

Demographic and health indicators in Niger are surprisingly stable. The only significant change since 1960 is reductions in mortality (Faulkingham and Thorbahn 1975; DHS 1992). In terms of well-being, Niger displays some of the worst indicators in the world: high mortality, poor access to clean water, and little or no food security. The Sahelian village remains a very hazardous place to live for everyone, but especially for women and children. The major causes of maternal mortality—obstructed labor, hemorrhage, infection, and toxemia—are all a direct result of the conditions of women's lives in rural villages (Boye et al. 1991). The conditions for children are hardly better. The village-specific birth rate for Guidan Wari is 52 per 1,000,[5] and 21.3 percent of the population is under the age of five (Jones 1995). Breast-feeding is nearly universal, and children are weaned at about eighteen months. The age group most at risk for malnutrition is between six months and two years. Iron and vitamin A deficiencies are common, as are respiratory infections, parasites, diarrhea, malaria, measles, and meningitis. Traditional herbal treatments for various diseases and wounds are common, and children continue to be bled as a treatment against illness (Jones 1995).

We now understand that mobility played a historical role of dampening population growth through out-migration and seasonal movement to other areas. The results of measurements of intercensal population change in Mayahi *arrondissement* show that between 1977 and 1988, annual population growth in the northern agro-ecological zones ranged from negative 1.85 percent to 2.35 percent, whereas in the southern zones, with greater agro-ecological potential, the rates ranged from 4 to 7 percent. Villages with fewer than 200 persons grew at rates averaging 1.31 percent per year, and villages smaller than 200 persons grew at a rate of negative 1.62 percent.

Has migration—either seasonal or permanent—in Guidan Wari had a marked effect on these observed demographic indicators? According to Ronald Lesthaeghe, male labor migration in Sahelian West Africa is continuing to produce traditional responses, which he characterizes as a heavy reliance by women on their children as their major productive assets, short child-spacing intervals due to the erosion of post-partum abstinence, the extension of childbearing into the mother's forties, and high rates of polygyny. Hence, Lesthaeghe reasons, levels of contraception are low (Lesthaeghe 1986).

But this is a circular argument, whose points are impossible to verify outside of the region where they are observed. The reasoning—that high levels of male migration beget a shortage of men, which begets high levels of polygyny, which beget spousal competition and high fertility—posits migration as an impediment to a successful transition to lower fertility. But it is a region-specific configuration, and no causation can be found without the heavy hand of functionalist social science theorizing.

Demographically, the villages of Guidan Wari are similar to many rural peripheries throughout the world. They are populated by old people and abandoned spouses, with few young adult men, and there is a palpable sense of severely limited opportunities. Though the effects are blurred by seasonality, circular mobility has raised the stakes for the rural population's ability to adapt. Seasonal absence imposes no inherent form of control on reproduction in this region where family size is "up to God." Mobility cannot be excised from the production system of the Sahel any more than the seasonal constraint of rainfall can. Imagining how the region would look without mobility is a useless exercise. To understand demographic dimensions, we must go beyond biostatistics to discover the importance of relationships. Villages do not live in isolation anymore. The next section investigates the social consequences of circular migration in the rural village, particularly for women and the elderly.

Social Consequences of Circular Migration

Community-level perspectives on the consequences of circular migration depict conflicts in the most basic of societal divisions: men versus women, and young versus old. The feelings expressed by the villagers who have remained—those who for various reasons are immobile—reflect the loss of conviviality and cohesion and the weakening of rural solidarity. Perhaps the most stark societal divisions occur along gender lines.

Circular mobility is an activity mainly practiced by men in Hausaland. However, it should not be forgotten that many women do migrate, with or without their husbands and children. In my study of circular migrants in Maradi, 39.1 percent of wives accompanied their husbands, and 22 percent more joined them later. As the villagers maintained, this *cin rani* mobility type is not necessarily forced by dire circumstances. Villagers recall the story of the husband and wife from Guidan Wari who went to Nigeria each year to work, and returned after the dry season fatter and healthier than they would have been if they had stayed home. Though the numbers of women traveling with their husbands is high, still almost 58 percent of the male migrants arriving in the city for the first time came alone; 38 percent of these solitary arrivals were later rejoined by their wives in Maradi. So about 22 percent of the male migrants left their spouses back home.

In his fieldwork among Hausa settlers in Tanout *arrondissement*, Jim Delehanty identified a mobility type that villagers called *balagaro*. It parallels *yawon Dandi* in Olofson's typology[6] and is a kind of opportunistic wandering when one's labor is not absolutely required at home. Delehanty describes the vein of humor underlying the seriousness of *balagaro*:

A Hausa man has the contradictory duties of guarding the women in the household but also of feeding one's household. The decision to go away carries a risk in the realm of the former duty, and everybody knows it, including the women in the household. Relations between Hausa men and women are much about sex and control over sex. It is a subject of deep seriousness and therefore also a subject of deep humor and joking (Delehanty personal correspondence, 1994).

By removing the titular head of household and normative decision-maker from the village for six to ten months out of the year, male migration can be empowering for women. This predicament is the subject of humor relating to the activities of "busy old men" in the village while the younger men are off on seasonal migration and the inexplicable rises in birth rates nine months after the period when husbands were absent. Gender relations transformed by circular mobility are also the cause of much resentment, stereotyping, and mutual suspicion between the sexes. Only very rarely do village women have the privilege of being able to go away by themselves without suffering severe consequences. Women are permitted neither by law nor by social norms or family exigencies to go off and seek their fortunes the way men can, unless they accompany their husbands. With the privilege of social power that is expressed through the mechanism of "remote control," a man effectively concedes what he perceives to be the small daily decisions of running the household to his wife or his family with whom she lives, while still reserving the major decisions—often the ones pertaining to land, livestock, and grain sales—for himself. In Rosalind David's (1995) case study from Senegal, the wife of one seasonal migrant in Dakar had to write him to get his permission to sell a goat. Men make the most important decisions, especially those related to children's education and marriage arrangements, from afar. Ownership of assets is an issue with blatantly inequitable gender overtones, and one that is critical to the efforts of development agencies and other organizations to promote social change in the village.

While I was in Guidan Wari, I interviewed a small group of young men, including Ayouba, a friend of Emlyn, in a shady spot near the Guidan Wari market. When asked about their views on circular migration in Guidan Wari, they described with great admiration the enterprising migrants from the village. These are successful men who proved themselves and gained a fortune in the world. Emlyn discounted the stories, and in the course of translating our conversation, she commented that these men might feel too proud to admit that things are so dire economically, that they and their families are hungry. She said, "Everybody knows somebody who can claim success, but how many are doing this themselves?" I asked the men if they feel that being away for five to six months isn't hard on their families. "*Ba komi, ba wahalla* [no big deal]," the men replied. "Besides, it saves food."

Their wives do not agree. Emlyn later told me that the wife of one of her assistants at the clinic used to be married to a migrant, but he never came home and never sent any money, so she asked for a divorce. But even though their four children were young, he and his extended family claimed and received custody of them.[7] Stories of women who abandon their villages and leave for the city, only to encounter their shocked and embarrassed husbands, are common gossip around the village. Women and men view the sexual transgressions brought on by mobility with a mixture of tolerance, defiance, and tongue-clucking disapproval. Women are entitled to half a man's share of inheritances, and the ownership of assets, fields, and livestock is normally in the husband's name or under the husband's family's control. However unfair this is, women have few options other than leaving, and even leaving is simply not an option women can take without severe social sanctions.

Migration plays a role in the timing of marriage ceremonies. The obligations to give money at ceremonies prompts men to sell their belongings and pick up sporadic labor in the city or in other villages so they can afford the obligatory gift, which can often cost several months' earnings. A man will often marry a second wife in May, prior to the rainy season, so he can receive the gifts and the labor she provides, since women work in the fields across northern Hausaland, from Tahoua to Zinder, and their labor is valuable. May therefore is a common time for marriages. This practice is partially made easier by government policy. In 1977, President Kountché set the maximum bride-price for a never-married woman at 50,000 FCFA (Boye et al. 1991), allowing husbands to recoup their investment from cash given as wedding gifts.

On the subject of marital relations, the cherished myths about households as unitary decisionmakers come crashing down. In a literal interpretation of Islamic societal norms about male and female roles, social norms are not rules that must be enforced. Islamic law (*sharia*) requires the head of household to "feed his family," but this obligation is sometimes interpreted by the man to mean leaving behind a sack of millet to suffice during his absence, however long that may be. Frequently the men leave without giving the family any money for *kayan mia* (sauce ingredients), and unless the women can find other food from the rest of the village, they are forced to drink *hura* (millet porridge) every day. Norms and actual behavior are rarely equivalent. Men are not obligated to buy clothing if they claim to have no money.

A male informant in Maradi described the gender differences between the village and the city. In the village, he said,

Photo 7.2   At a Guidan Wari well, women support the continuation of the rural village existence through sheer will.

the woman's life is harder, because she has to work so hard. Moving away from the village and living in the city is a remedy for this. But in the city, the man's life is harder. In the city, men do traditional female occupations; they wash dishes and even pound millet, whatever odd job that would be beneath them in the village, and no one from the village need ever know. In the village, women work from before sunrise until late at night—pounding, getting water from the well, getting wood, and cooking. The money they make from raising livestock and growing food in their own plots is sometimes taken by the husband to pay gambling debts.

The commonness of this experience was verified by others. Men are distant from family life even when they are home in the village. Often they do not even eat in the same compound as their families, preferring to spend the family's cash on prepared street food. There seems to be no place to hide anything in the village, yet men consort with prostitutes, play cards, and drink tea all day beneath the large *gao* tree in the center of Guidan Wari, while their wives proceed through their daily trips to the well (Photo 7.2), pounding millet, preparing food, feeding animals, disciplining children, and doing countless other tasks that fill their days. Given the men's disengagement from the daily affairs of the household, it is no surprise that women are not emotionally devastated when their husbands decide to live outside the village for part of the year. In many villages, it makes little difference whether the men are at home or in Maradi, except that while the men are at home, their wives may bring the lack of food in the house to their attention.

However bleak this situation might appear to outsiders' eyes, women gain a measure of autonomy from their husbands' absence, and there are indications that their secondary status may soon be challenged. Overt resistance in the form of political mobilization has as yet registered few gains. The opposition to *Le Code Familial*[8] by Islamic groups in the early 1990s was fierce and widespread. Women demonstrated in favor of democracy in 1991, and again in opposition to the jailing of political candidates in 1996. Mobilizing political support is often impeded by social class obstacles. The national women's organization, Association des Femmes, is dominated largely by wealthy urban middle-class women, and it tends to overlook the needs and circumstances of the rural poor.

Covert resistance in various forms, what James Scott (1985) calls "weapons of the weak," is far more common. A woman learns to play the decisionmaking game, leaving visible large decisions to her husband to satisfy his need to control, while quietly supporting or undermining him. Women act in solidarity with other women in the village, sharing food and labor with other households, cooking for each other, and lending each other their children for work. One benefit of cloistering, or female seclusion, for rural women is that the restricted access by men enhances female solidarity, which affords women the chance to operate in a female-only world. Empowerment through Islam can be a positive reality that Western outsiders rarely see. When nothing else works, women simply cope, making do with what they have.

Thus, an unintended consequence of circular migration is women's acquisition of the power to run things in the village, even in a hidden way, and this is something that women are just now beginning to exploit. In Guidan Wari, women have started a community garden near the well, having bought their own seeds from the Peace Corps volunteer. The women want to sell tomatoes and eggplant, but without better connections to the outside, there is only a weak market outlet.

"The outside" has only now begun to respond to the needs of Guidan Wari's women. The Bamako Initiative, which is funded by the United Nations, has begun to improve child and maternal health and family planning services for the women in Guidan Wari. For the first time, women may get contraceptives without their husbands' permission, providing the women with a chance to "rest" after having a child. Without this, women are exposed to the unfortunate coincidence of the agricultural and gestational calendars, with respect to sexual contact with their husbands. The husbands see their seasonal migration as being necessary

Photo 7.3   *Peace Corps volunteer Emlyn Jones shows her storyboard with the example of two families. The one on the left practiced child spacing and thus was able to produce hardier children, whereas the one on the right did not and thus suffered higher mortality. Family planning mostly remains just a good idea in the rural parts of the Sahel.*

to the functioning of the household, but they do not want it to affect relations with their wives. The result can be a ceaseless annual childbirth or a debilitating series of health problems or both.

In a rural village five kilometers from the nearest road, a white Peace Corps volunteer gives health lessons in fluent Hausa—two separate lessons, one to the gathered women and one to the men. In each, the volunteer illustrates a story using a large, handheld storyboard that pictures two families (Photo 7.3). On the right are a father, mother, and two children, looking tired and unhappy, with drawn faces and wearing old clothes. On the left is a father and mother along with four children ranging in age from toddler to teenager. The volunteer explains that the family on the right intended to have a large family, as is the norm. But they had too many too fast. When a younger child would come along, the older one still wanted to nurse but had to stop. The two-year-old weakened and died when a measles epidemic came to the village. In the family on the left, with the four children, the mother used oral contraceptives to "rest" after each child. As a result, each child could nurse for two years, and more of the family's children survived. Now the family on the left has a large family, which they were able to have thanks to family planning.

The idea that by spacing children's births out more, the mother will have more time to attend to each child and mortality will thus be reduced is very understandable to the audience. The villagers in both groups nod and mutter words of comprehension; even the men see the logic behind the idea of "resting." It should not be a surprise that only a pro-natalist argument works effectively in this setting; it appeals to the desires for a large family in order to assist with fertility reduction efforts.

This intervention is technologically and culturally appropriate, and as the sketch showed, it is locally understandable and rational. But at the same time it stands little chance of changing the realities of village life substantially unless it addresses the underlying realities and contradictions of Sahelian rural life. Women's livelihoods need to be supported and made viable. Attention needs to be paid to the essential inequalities in the rural male-female relationship. Decisionmaking is currently in the hands of Allah. How would the villagers' lives be different if they claimed power over their own destiny? Technological and environmental constraints must be addressed. Simply promoting contraceptives to the rural sector will not in itself change anything.

In the end, African society is built upon the backs of women, and the reproduction of the village existence eventually comes down to relations between the sexes, or lack thereof. In their mobility, men have an outlet that most women—because of culture, responsibilities, and obligations—do not. This differential freedom revolving around gender is culturally sanctioned, and it is reinforced by years spent both together and apart. At the moment, there is very little that women can do about it, but this can and will change. The vicious circle of high fertility and mortality in rural-based peripheral economies will eventually become antiquated as urbanization proceeds. Women exposed to alternative lifestyles and modeled behavior in the city will choose autonomy over patriarchal control, which continues to be associated with rural areas.

Like conflict over mobility and gender, a similar dynamic of age threatens to split the village society apart. The conflict pits mobile young men against the less mobile older segments of the population. Old people wish to preserve what they perceive to be cultural coherence, whereas the young want to be free to pursue opportunities elsewhere. The young men in the village of Guidan Wari claim that they do not go away nearly as much as men in smaller surrounding villages do because the old people in Guidan Wari shame them into staying home. The young men say, "*Their* parents didn't go away, so they do not want their kids to go."

Illa, the old village chief, holds strong views on the subject of migration. He knows that money shortages are at the root of young men's desire to go away. Migration is so obviously bad, according to him, that it is destroying the village. But, I respond, hasn't this mobility always occurred here in various forms? No, Illa says with his finger wagging, "When I was younger, boys herded animals and did not have time to go away like this." In other words, they still went away, but perhaps not as far. In an aside to me, Emlyn explains that it is in Illa's political interest to keep people here. New ideas find their way in, and the old ways are challenged and discarded. In the view of the gerontocracy, mobility challenges the chief's power and worsens divisions within households.

Apparently this old-fashioned isolated rural village is coming apart at the seams. As the center of self-interested deliberate stasis and the preservation of status quo, the rural village has been painted as a locus of surveillance and dominating social pressures (McNicoll and Cain 1990). It is tempting to draw the village as a paper-thin movie set, maintained affectionately by its former residents. Yet the village is still where the bulk of the country's food is grown, and it continues to be a source of cultural values and norms associated not only with the negative aspects of Sahelian existence but also with the positive ones. Banding together and resisting a common foe, distributing community resources equitably, and striving for the survival of the less fortunate are all abilities that were forged originally and continue to be reproduced in the rural sector.

However, just because rural villages possess the solidarity to survive droughts, epidemics, and famines does not mean that every village member is treated the same way. Though it is usually not intentional, the neglect of children is an age-old response in times of drought or other crises. The Sahelian village is an unlikely laboratory for experiments in social Darwinism, but every time a child's fate is put in the hands of Allah in this pro-natalist society, his or her parents are following an unspoken conviction that the child is ultimately replaceable. Viewed through the eyes of children, the hypocrisy of drought-time famine-mitigation practices calls into question the intentions of adult stakeholders. Was village solidarity ever as well-knit as when the strong who survived the epidemics put themselves first in line to eat, counted on gifts of food from faraway friends or dependents, and could survey the devastation and vow to continue? Few villages are self-contained entities that feed themselves and persist in excluding the outside world. When the gerontocracy is deprived of the privilege of being survivors and instead must share responsibility for the future—which includes caring for the children—will they proclaim victory or concede loss?

Discussion: Poverty, Mobility, and Environment

The urban view, expressed by migrants, urban merchants, and the tiny minority of formally educated teachers and civil servants, asserts that the village as a viable cultural source is obsolete and fictive and that it is driven by the affectionate remembrances of former residents. Rural villagers have ample reason to challenge this characterization. Yet villages today fit into a socioeconomic network whose span reaches from the individual investor or peasant to an entire global economic system. Labor circulation is but one of the manifestations of the connectivity of a village with the wider network, serving to strengthen the wiring that supplies the ideas and increasingly the resources.

Who in fact is more vulnerable: those who migrate, or those who cannot? Throughout the study we have seen that increasing the connections with the world beyond the village, with its opportunities and markets, provides benefits that are abstract as well as material. Socially, kin and affective ties are flexible, enduring, and most useful for surviving droughts, famines, and crunch periods. Economically, having better access to news outside the village allows for a diversification of interests and investments to spread risk around and promote long-term stability. Being able to use land and other local resources allows people to capitalize on variability, it guarantees a food supply, and it is more environmentally sustainable. Moreover, mobility is an indigenous solution to an otherwise stubborn problem: how to persist, and even succeed, when nature has dealt a lousy hand. But it is hard to see the advantages of mobility for those who, for reasons of culture or obligation, are unable to leave.

Class and the local power structure in the village are difficult for an outsider to discern. In Guidan Wari, where the economy of affection and obligation still thrives, a landlord class strictly speaking is absent, and yet a notion of social class or caste has manifestations that are visible in the various types of mobility practiced. Is there in fact marginality within this Hausa-Bougajé village? Differentiation has not reached this community in a way that is discernible to outsiders; noticeable gaps do not exist between what we in the industrial West would call the "well-off" and the "poor." During a drought, the ones with food will share it with those who have none, and the less fortunate patiently hope they will not have to beg. And in most cases, there are no convenient villains. At the time of my village interviews, all the children of one of Guidan Wari's *el hajis* were underfed, and one of the chief's own grandchildren was gravely malnourished and was expected to die soon. In access to medical services, beliefs about disease or nutrition, or fatalism, there is a level playing field in this rural village.

Overlooking for a moment the ruptures and social divisions, a functioning rural society here can be imagined. The villages surrounding Guidan Wari cannot grow outward as they always were able to in the past, but perhaps they can pull themselves up with a small-scale metropolitanization. Market growth would come not necessarily from the city of Maradi, which is a day's drive away by local transportation, but it could come from Guidan Wari itself. There are about twenty-two villages and 3,500 people are within a five-kilometer radius from Guidan Wari, which translates into walking distance of an hour or two. Guidan Wari has a clinic, a working weekly market, and daily or almost daily truck service to Maradi. Given the environment—the seasonality of rainfall, the inevitability of drought, the poverty of soils, and the scarcity of year-round water sources—there is not much to cling to, but the relationship between Guidan Wari and smaller villages must improve.[9] The smaller villages cannot be self-sufficient anymore, and neither can Guidan Wari.

Therefore, we must break with the conclusions reached by other outside observers who argue that migration provides short-term need satisfaction but has negative long-term consequences and that ultimately the costs of migration are higher than the benefits (Amin 1974a; Painter 1987; Cordell, Gregory, and Piché 1996). Cordell, Gregory, and Piché write: "It is clear to us that circular migrant systems breed poverty and underdevelopment" (p. 317), and, "the general consensus is that remittances have had little impact on development, because they have not led to new productive activities." Such sentiments are shared by those who seek to return the marginal region of the Sahel to an idyllic and largely imaginary past, a past in which villages held together in hard times and cultural coherence was preserved. Today's migrants, as well as those who are left behind in the villages, do not have the luxury of such categorical statements. Given the changes in the world and in the relational attributes between rural and urban, isolation is no longer an attractive option.

The role of mobility in enhancing drought preparedness should not be ignored. What does return to the village travels under the guise of family obligations and the enduring idea of household food sufficiency. When the next big drought hits, as it always will, having exit options will ensure the survival of some. And circular and seasonal migration helps exercise the social networks, allowing individuals and families to keep their options open. Mobility is the medium of transition toward sustainability, and it is the people's most effective existing means of making decisions.

Niger will never again be a lush and productive land unless people allocate their time and energy differently. They won't, because their allegiance is to their people, not to their spot on the map. A village is a community of shared interests that is only partly rooted in a established physical location. "African space," Kopytoff writes, "is above all social space" (Kopytoff 1987, p. 22). The village is not a collection of structures but a living organism with a common history and a common purpose, which can and does move until space no longer permits it. The implications of this are that those who can escape the village will, and the village will be remembered fondly as a source of Nigérien culture—less viable today as an economic unit—but still supported by its contacts with former residents.

In this chapter on community responses to and perspectives on circular mobility within the rural sector, I have shown that the site-specific erosion of self-sufficiency creates a characteristic map of seasonal circular population movements. In the drier, northern zones, hunger mobility of the temporarily displaced is more common, whereas the wetter south is marked by investments in food production and the circular movements of resources. The economy of Maradi has been realigned toward occupational specialization, with circular migration playing a leading role. The grand process of urbanization occurring in sub-Saharan Africa affects people in different places differently. Where does the rural village fit into these changes? Is Maradi's brand of short-distance and network-active migration truly a two-way street for the rural sector?

In geographic space, there appears to be a clear line, represented best in areas of similar agro-ecological potential, where investments in agriculture are perceived as being too risky and the hazard of drought is mitigated by spatial mobility. As we have seen and as the migrants themselves have documented, the practices of seasonal circular migration relate less to environmental degradation than to the process of landscape domestication. Mobility has always been a feature of the Sahelian zone, and the transformations regarding the erosion of self-sufficiency and the growing reach of the city parallel each other. A durable solution to the problem of resource endowments, and one for which Hausa peasants have individual and collective experience, is the diversification of risk. At the moment this strategy appears to favor population centers.

Notes

1. Villagers were asked to derive the levels of male mobility by type, by estimating the number of men *out of ten* who practiced each type. I then rounded the numbers into a percent. This provides a very rough picture, but the best that could be had given the circumstances.

2. To confirm these black market prices, I asked in the Maradi market, where I was told the price had been 2,500 FCFA in 1994 for a fifty-kilogram sack, which was 5,800 FCFA when I asked in 1996. It is difficult to separate the price rises owing due to increased black market demand and those associated with the loss of fertilizer subsidies. Clearly, the demand for chemical fertilizer has exceeded the supply available from the government.

3. John Curry (1989) describes some traditional responses to drought, including flexible local production strategies, exchange and market networks, food sharing and other forms of risk insurance, dietary diversification, food preservation, dietary exchange, agricultural innovation, social organization, and increased participation in the market economy. These responses are, I feel, not di-

rect responses to drought but responses to *conditions* in which drought plays a role.

4. The map was produced with the help of a group of Islamic scribes in Guidan Wari, who identified households from a sketch that I made of the village. I also received help from Kalla, the town crier, who identified households whose "sons" were not year-round residents of the village. The results were then confirmed with the younger chief and his retinue.

5. This birth rate is comparable to that of Niger overall.

6. *Yawon Dandi* (the walk of Dandi) is described by Olofson as "an aimless, pointless wandering" (Olofson 1985, p. 55).

7. Norms about marriage and divorce in the Sahel vary by region and ethnic group and stem from a blend of customary and French colonial law. Among the Hausa, a husband can "repudiate" marriage—dissolve it unilaterally—and obtain custody of the children, especially older ones (Boye et al. 1991).

8. *Le Code Familial* (Family Code) is a statute best defined as a government effort to regulate traditional marriage and childbearing practices.

9. In her study of the rural Hausa in Nigeria published in 1972, Polly Hill made a similar argument for small-scale metropolitanization. She described the causes of rural poverty as including "the brevity of the farming season, unreliability of climate, under-utilization of labor reserves during the farming season, shortage of working capital, shortage of cattle manure, the dearth of remunerative non-farming occupations during the dry season, the inability of poorer men to finance migration for farming, village 'balance of payments,' difficulties owing to small range of produce and crafts goods which are sold 'abroad,' and the burden of assisting poverty-stricken people, which is borne by the community generally" (Hill 1972, p. 190).

8

Walking into the
Next Century

This book has examined the demographic, environmental, and political-economic contexts of an urbanizing peasantry in the West African Sahel. By showing the transformation of circular mobility in both the rural and urban settings, this has been a case study of the interactions between population and environment. *Cin rani*, along with other Hausa archetypes of circular seasonal migration, is a cultural form eminently suited to the seasonal precipitation regime of the Sahel; it is also a form that has found new life in recent decades. In the story of Maradi's transformation as a region, villagers have come to rely increasingly upon mobility—a normative, culturally sanctioned activity—to satisfy food and cash needs. I have argued that this reliance cannot be separated from the process of settlement and the methods of production in the places where migrants originate.

As is all human behavior, mobility is locally constructed and rooted in both social and physical environments. Along the environmental margin, the seasonality of precipitation and drought risk prevent year-round rain-fed cropping. Shifting cultivation is used to maintain a balance between food needs and growing conditions, and where sustained yields cannot be assured, food needs must be satisfied elsewhere. I have suggested that in the marginal zone of the Sahel, which was settled mainly in the twentieth century, distance and space have played central roles in gaining food security. Mobility systems have transformed in response to local changes, such as increases in population density and disappearance of bush land, following a general path. As people's village-based options for food security shrink, forms of seasonal migration have changed from those devised to exploit the frontier—including herding, hunting, and off-season movement to other agricultural opportunities—to those that utilize portable skills to seize opportunities now located in the city.

A major objective of my research has been to relate such local and regional changes to the individuals participating in them, thereby achieving a kind of "grounded" geography of human population along the environmental margin. The attempt to link migration and environment reveals some obstacles to achieving a unity of design. In the standard approaches to migration, worsening rural conditions are considered "push factors." This stimulus-response calculus ignores individual and group strategies and renders deterministic the decisionmaking process. Though depictions of "environmental refugees" are in some cases polemical and overgeneralized and though the environmental influence on population is sometimes nebulous, the forces that operate to root people to, or uproot people from, their places and routines cannot be so easily dismissed. Orthodox migration approaches are useful for isolating determinants in a cross-sectional sample designed to be representative nationally, but there is an insufficient amount of geography in these approaches to show how social and environmental contexts interact at the level of human settlement.

Linking the "push factors" to existing local environmental conditions requires a better understanding of place-based history and geography. Ways of adapting to change are built partly upon culture: that is, time- and space-specific accumulated actions springing from the interactions between population and environment. By its nature, culture is place-embedding: the sum of actions and consequences of actions occurring continuously in the same place, *or among the same people*, over time. The task of drawing culture and mobility together is especially challenging in regions where the neoclassical model of migration, in which behavior is influenced by calculations of anticipated income, can explain why people move but not the reasons for mobility itself. Accounts of forced population movements in history normally assume an ambient prior level of sedentarism, and the victims of famine are normally assumed to have been rooted in place prior to the events that uprooted them. In non-Western settings—among Inuit seal hunters, Bolivian highlanders, Mongolian plainspeople, Hmong forest dwellers, and Hausa millet farmers—forms of mobility, especially circular ones, are de facto components of production systems. Urbanization and modernity influence these forms, altering but not erasing them. The relationship between changes in the environment and in mobility practices over time requires a more serious and comprehensive treatment.

If the Sahel as a region is defined in part by the mobilities practiced there, then outsiders' blindness to the uses of mobility, compounded by an emphasis on permanent migration in demographic theory, influences the potential for development along the environmental margin. Migration is often interpreted as an indicator of conditions that are under social or environmental stress. It is presumed that if villagers were only provided with the right conditions, villagers they "set down" and invest their efforts in more locationally fixed production systems. In theory, this assumption makes sense; in practice, it is often ignorant of the geography of the marginal environment. As evidenced in the failure of the FAO-funded showcase development project in Tahoua Department, as long as a severe drought occurs every five or ten years, natural resource management and food-for-work programs are ineffective as glue to stick mobile peasants in their villages. There should be no surprise that such development approaches have failed in the Sahel, a macroeconomic climate that squeezes small producers and penalizes peasants for acting in their own interests.

As strategies of adaptation that use mobility to overcome physical limits to food security, the choices of livelihoods themselves represent responses to environmental change. This study has illustrated the process by which rural peasants become connected to an urban social network of family and friends to find housing and employment. Through access to cash and opportunities in the city, migrants are able to support their families back in the village, socially if not financially, and maintain their viability as food producers in the rainy season. With reference to the question of whether remittances from migration are financing agricultural change in the rural sector, this study found that significant numbers of migrants did send cash, food, and other items back to the village. Whether migration serves as an effective mechanism for rural development, however, is unclear. This is especially true in the case of agricultural investment, whose rationale is obscured by cultural differences. The Western model of capital investment as a kind of anonymous gaming parlor is inappropriate for the setting in question. In West Africa, investment means alliance, in which obligations and tribute—the mutual interdependence of an employer and employee and the ties that bind families together through marriage—allow individuals to work the system to fulfill needs and gain a measure of security. If the risk of crop failure is too great to permit a reliable yield, peasants will choose to honor their web of locationally diversified social ties to satisfy food requirements.[1] I have argued that these decisions, and the uses of mobility to implement them, spring from the environments in which the social structures are rooted. The livelihoods cannot be forced back into settings where they no longer function.

The environment may have a clear effect on population mobility, and how the environment in question is defined will obviously influence the prevalence of population mobility (that is, how many people go away and the manner with which they move). Without knowing how long people have lived in a particular place or how labor and land uses interact and without understanding the strategies employed by individuals to improvise and utilize livelihoods to adapt to changing circumstances over time, all the geo-referenced resource inventories in the world will not show how people derive meaning from the places they inhabit or how people regard opportunities elsewhere. To achieve a grounded theory in population geography, there must be a mechanism to capture the dynamic of movement and also to account for place, culture, and environment. This cannot be done unless migrants are first presented as people, and not just as populations.

Population geography is in sore need of human faces and voices. This study is set in contrast to approaches that transform people into faceless sets of numbers; it has taken seriously the idea that people participate in the changes that occur around them. Tap an urban migrant on the shoulder in a West African city and you will meet someone with a story. By listening to and recording people's stories, the importance of such nonmaterial resources as ingenuity is revealed. Not only that, but a geography is revealed as well, and the challenge is to link individuals to the larger scene. Livelihoods have a particular logic that benefits from a geographical approach. Since they are rarely "mapped" or otherwise quantified, livelihoods tend to be omitted from consideration or undervalued.

A grounded population geography must link people with their environment in order to present a coherent alternative to Thomas Malthus and his dehumanizing image of miserable peasants. Ester Boserup theorized the positive effects that human populations can have in changing the conditions of agricultural growth. Population clearly plays a role in environmental decisionmaking, for land uses vary along a continuum of population densities. Yet there is a tendency to regard population density itself as the cause of agricultural change, when in fact such a measure is only a reflection of broader forces of societal and environmental change.

In this study I have argued that population mobility plays a significant role in agricultural land-use dynamics, making Boserup's position more geographical by theorizing mobility *into* a diversity of changing production conditions. Where the transforming effects of higher population densities cannot make themselves felt because of environmental limits, communities often find themselves caught in a bind. They have increasing needs for food but are not able to invest additional labor in land when conditions are so risky. In such circumstances, seasonal population mobility emerges as a step in the density-activity relationship to fill the gap between production capabilities and consumption needs. Villagers who move about seasonally are gaining greater access to the world of opportunities, which will ultimately have consequences for their villages of origin.

Even in Niger, one of the poorest countries on earth, the pattern of relations between people and their environments is persistently complex.

This complexity is revealed in several ways. In this study, population mobility is a demographic variable, but it is also a resource, an economic strategy, and a value system. Winners make efficient use of their mobility, as well as the mobility of others, whereas losers are fixed in place, the captives of misplaced investment and cultural norms that undervalue their efforts. Mobility lies at the crossroads of the relationships between people and environment, structuring how economies work regionally and how communities respond to hazards. The experiential links between settlement, land use, and economy can be understood as they interact through mobility. How people move is as significant as how they use land, if not more so in cases. Circular seasonal mobility may not be the engine for rural development in marginal places, but because it allows social ties that span space to be maintained—and even strengthened—through seasonal ebb and flow, mobility makes people better prepared for hazards and ultimately reduces vulnerability.

Regional Changes in the Study Area

After introducing the study region and the theories of West African migration, the book undertook an examination of the geographical context for seasonal circular mobility in southern Niger. In Chapter 2, I examined the context of population and settlement dynamics, based on the contention that one cannot understand how land use affects population mobility without returning to the conditions prior to and during initial settlement. For the purposes of the study, I defined the region of Maradi by historical settlement patterns and land uses, both of which use mobility. Settlement patterns in Maradi have been determined by the presence of water. Most people live in the southern agro-ecological zones, where levels of precipitation are high enough to permit crop agriculture during most years.

Starting in the 1800s and accelerating in this century with the French pacification of the region, Mayahi in the Sahelian zone received immigrants from other places. From the north came Tuareg and Bougajé nomads and seminomads. From the south came peasants from ancient Hausa states, fleeing tyrannical rule by the Fulani zealots of Sokoto. The frontier era was marked by hunting and herding mobility, as well as by Hausa trade movements, including *fatauci* (long-distance trade). Land uses in the Sahelian zone were characterized by a system of shifting cultivation of millet and sorghum and by the use of bush fallowing to restore soil fertility. Security in the area was assured by French military occupation and civil infrastructure. Additional factors—such as a wetter-than-average climate, freely available land, the French encouragement of groundnut production, and an active frontier ethic—encouraged the filling of the Sahelian agro-ecological zone through the 1920s and 1930s. The French introduced a road network and other infrastructural improvements into formerly isolated rural villages, and basic public health measures helped control epidemics and reduce mortality. Following the closure of the agricultural frontier in the early 1960s, the system of shifting agricultural production came to be challenged by population growth through natural increase.

Transitions in population mobility were built atop these preexisting conditions of regional development and population. The relationship between settlement dynamics and present-day mobility is visible at the landscape level. The temporal order of village formation affects villagers' access to resources: Older villages have better locations with respect to wells, trees, and farmland. Newer villages built between those settled earlier have poorer access to resources, and they tend to be the villages where seasonal migration is most prevalent. Embodying periodic movements to better and more abundant resources and the continued reliance on distant affiliations and loyalties, seasonal migration is an extension of age-old practices, now accelerated by the endogenous process of frontier settlement and the disappearance of bush land. Local settlement dynamics cannot be viewed solely in microscale isolation; rather, it must be placed in a larger regional context of urbanization. Most of the movements in rural Maradi today are short-distance rural-urban mobilities, directed toward the department capital or cities in northern Nigeria; their objective to engage in commerce or to seek sporadic short-term employment. The transition from exploiting the frontier to employing circular seasonal mobility could not have occurred without the organizing capability of the city.

The second context that I studied was that of land-use intensity change. Chapter 3 extended the historical-geographical focus of changing conditions in Mayahi *arrondissement*, in the Sahelian part of Maradi Department, to look specifically at the role of land availability on the ability to grow food. The theory of induced agricultural intensification as stated by Boserup and others is that land-use intensity affects available options for food provision. Boserupian explanations describe the effects of rising population densities on production systems, including the reduction in fallow periods, the increased investment in land, the shift to animal traction, the increased soil fertility maintenance through use of manure and fertilizers, and the change to individual land tenure arrangements.

According to Boserupian theory, land must be scarce before production changes take place. In Guidan Wari, mobility appears to be influenced not just by population density and environmental conditions but also by the common perception that fields are incapable of providing sufficient nutrition. Interviews in the study villages found a clear connection between the closure of the agricultural frontier and an increased prevalence of seasonal migration. Villagers realized the potential for improving production, but they were unwilling or unable to make the necessary efforts. Seasonal migration was therefore rediscovered as a response to environmental change, as the old ways of adapting through frontier movements and land conversion no longer worked.

To determine the precise dimensions of the changes in land-use intensity from 1975 to 1995, an analysis was done using remotely sensed land-use intensity data, including high-resolution aerial photos and digital videography. It was clear through the comparison of changes in land-use intensity that change varies by agro-ecological category. Wetter land with less compacted soil was often already filled by 1975. After 1975, land cover in the drier and more denuded areas of the north was converted to agriculture. Video footage from 1995 reveals that even in the drier north, farmers have put many fallow fields and bush areas containing stands of trees into production, and even barren hillsides and gravel-covered seasonal riverbeds are being exploited for crop agriculture. In the driest and most remote regions, land is still available. Despite the apparent omnipresence of environmental limits, peasants from many places denied that farm land was scarce, suggesting instead that the causes of their food insecurity were insufficient rainfall, the lack of soil nutrients, or wind erosion.

The barrenness of present Sahelian landscape raises questions about the permanence of environmental changes. Over the past century, resource mining has reduced the productivity of the land. The removal of biomass has lessened the area's capacity to cope with erosion and to retain moisture through the dry season. Many villagers agree with the view that stripping nutrients impoverishes the land, and yet grasses and millet stalks that could be left in the fields to preserve moisture have a tangible economic value to households if removed. Given the north-south livestock routes and the tendency of ruminants to disperse seeds, much of today's brown landscape potentially *could* be greened. The preservation of certain tree species, woody shrubs, and wild foods with human value attest to the human control of the Sahelian environment. Even so, there is little to suggest that revegetation will occur here, for reasons having less to do with creeping deserts than with the timing and location of human economic activities. With insufficient labor for the extra inputs necessary and with the gearing of the agricultural calendar toward maximizing the time spent in the city earning cash, there is little practical incentive for intensifying agriculture to raise yields. The potential for agricultural production is severely limited by the seasonal absence of the most able-bodied farmers, who have allocated their labor elsewhere for much of the year. This may explain the villagers' perceptions that food shortages exist even though more land is still available for agriculture. Without considering the massive periodic relocations of labor and their effects on sustainable land use, there is no point in calling for increased conservation practices.

The third context that I studied was that of political-economic change in the Sahelian zone. Prior to the French conquest, the precolonial economy was based on age-old interactions between grain-growing sedentary farmers and livestock-herding nomadic pastoralists and on long-distance trade between the desert edge and large cities to the south. The history of indigenous markets in the Sahelian zone reveals a pattern of incorporation into a regional economy that accommodated the climate through the use of seasonal movements. As midnineteenth-century accounts confirm, cotton was grown in the Sahelian zone long before the French set foot in the area. This and other such desert-edge products as grain, skins, dates, and salt were traded with cities along the West African coast. The spatial configuration of markets and cities was not introduced by the French, though the imposition of colonial control, and its administrative and commercial mechanisms, made department cities more influential and authoritative.

The French promotion of groundnuts in the 1920s and 1930s allowed the Sahelian population to spread and to grow. The groundnut boom enabled infrastructure such as roads and marketing boards to be put in place, along with market outlets for later food sales. The colonial apparatus fostered the growth of an *el haji* monied class, which accumulated capital and state benefits and fed them into informal urban economies. Ties to the countryside through commerce increased throughout the colonial period, and this did not alter substantively after Niger's independence in 1960.

The political economy was disturbed profoundly by the series of droughts that occurred in Niger and throughout the Sahel starting in 1968. Hundreds of thousands of hungry migrants flocked to the cities, replicating an old practice. The distress sales of livestock and other assets raised fears of predation by the urban monied class. Niger's first coup d'état occurred in 1975, after seven years of below-average rainfall. In response to the drought, the state emerged to mediate between the bourgeoisie and the peasantry and to restore social order. Seyni Kountché's Société de Développement (Development Society), founded in 1975, sought to create a new nationalism through a publicly financed social movement involving quasi-traditional groups. This ultimately served the needs of the state during an era that was marked by a growing dependence on food and development aid. State coffers were also filled by revenues raised from uranium mining in the north of the country, a region formerly under the control of the nomadic Tuareg groups. The string of growing towns and villages along the national highway in the extreme south marked the increasing dependence on outside money for the function of the postcolonial political economy.

One legacy of Kountché's era was the attempt to regulate seasonal mobility through state control. Prior to the 1974 coup, the first president, Hamani Diori, had encouraged mobile education for pastoral people and better ties with France through international migration. After the coup, which deposed the First Republic over perceived excesses and corruption, the reformist Kountché saw seasonal migration as a cause of rural poverty and sought to control it through roadblocks and mandatory identity cards. Only after Kountché's death in 1987 was the utility of mobility acknowledged for rural villages. Remittances from international seasonal migrants on the West African coast were collected by migrants' associations and funneled to Nigérien villages to purchase food and other necessities. These resource flows also fed the growth of Tahoua and cities along the southern frontier such as Gaya, Dosso, Birni n'Konni, Maradi, and Magaria, which acted as gateways for short-duration movements to points further south.[2]

Given the current political-economic climate, the importance of mobility to the provision of nutritional needs cannot be overstated. Niger is a country with feet. The contexts of settlement dynamics, land-use change, and political-economic change operate simultaneously, and together they form a universal process of change in which mobility plays a key role. As villages used up their land and villagers were casting about for ways of replenishing stocks and gaining food security, the growing city came calling with needs for labor, food, cash, and markets for goods. These changes occurred in parallel, as a mutual reaching between the city and the village. From the 1950s to the 1980s, the slowly transforming situation was punctuated by droughts, which squeezed accumulated skills and assets out of villages and moved them to cities. As cities grew in size and importance, more and more daily and weekly movements between nearby villages and urban hubs occurred. In effect, the transition of mobility from a frontier-capturing mobility to a rural-urban circular mobility spelled the emergence of a *culture routière* (road culture) across southern Niger, utterly dependent on the road for the flow of resources.

I contend that what has occurred in Maradi, in the peopling of the environmental margin, has characteristics that can be found across the African continent south of the Sahara, especially in West Africa. Increases in seasonal migration are clearly the result of worsening conditions. It is tempting to ascribe the changes in mobility patterns to the combination of low resource values, meager agricultural inputs, abundant though not extremely productive land, and scattered nonagricultural opportunities.

All of these are common to sub-Saharan Africa, making hunger migration partially a consequence of the African environment itself. But this is not what is so African about African mobility. What gives African mobility its special status is the presence of an undeniably real challenge to human development posed by the current state of politics. Poor, weak, parasitic states without recognized authority, illogical and inefficient borders, and economic conditions that are often in direct contradiction to the will of the state, especially in informal-sector activity, are far more important than environmental change alone in explaining the increases in seasonal migration.

Seasonal migrants are clearly significant challenges to state authority. Postcolonial movements are popular expressions, carried out without the blessing of the state and often in opposition to the state. In the case of Maradi, a strong indigenous state apparatus has never existed, and people have defined themselves in opposition to the Sokoto zealots in precolonial times or to the Zarma-dominated Niger national government in postcolonial times. As far as I observed with reference to Maradi, the state was essentially meddlesome but ineffectual. In the handful of instances where the state appeared to play a role, such as the army crackdowns on the Maradi shantytown, the police stops outside the city (which the taxi and truck drivers avoid), and the seasonal migrants' visible fear of state tax agents, the opposition between migrants and the state is clear.

The recent turn of national political events in Niger only worsens the widespread cynicism in the Nigérien populace. After a split between political parties led to impasses in the democratically elected national government, the civilians were overthrown by an army coup in January 1996. The coup leader, Ibrahim Baré Mainassara, ran for elections in June of the same year, jailed political opponents, drew protests from every international observer save France, and won overwhelmingly (with disputed results). He made overtures to other dictators in West Africa, including Sani Abacha in Nigeria, and he welcomed the Libyan leader Mua'mmar Gadhafi, who landed four Boeing 727s in Niamey in defiance of UN flight prohibitions against Libya (*Camel Express Télématique*, May 10, 1997). As a result of Mainassara's questionable victory, the United States, Canada, and Switzerland pulled all nonessential aid out of Niger. In April 1999, Baré was assassinated by his own presidential guard, and the National Assembly was disbanded.

These incidents have cast a shadow over Niger's future. As conflicts in Somalia and elsewhere have illustrated, very little support exists for the idea that people can organize themselves into coherent units without some direct involvement by the state. Keeping the situation from destabilizing progressively will be very difficult, if countries like Liberia or

Sierra Leone are representative examples. In these unstable areas, young men with little or no experience outside the village are pushed together to wait like dry tinder to be ignited by a demagogue. The rise and fall of the *Maitatsine* millennial movement in northern Nigeria and the recurring persecution of fundamentalist sects in Maradi could presage this for Niger. However, outsiders must be wary of painting the situation with too broad a brush. Given the complexity of the changes occurring in Niger, a focus on people requires a more sensitive interpretation of conditions and events, and some diversity of responses is expected. *Are* the seasonal migrants in fact unstable elements, as they are sometimes portrayed? Where do they come from, and what do they do in the city? What are the relationships between their city existence and the villages they came from? To what extent are migrants forced into their circulatory habits?

It was with these questions in mind that I undertook a survey-based investigation of the changing worlds of the seasonal migrants in the city of Maradi. Viewed through the lenses of society, individual and community responses reveal the complexity of human agency. Mobility is both a necessity and a choice, and it is both a traditional and a modern behavior. Individually, seasonal migrants are overwhelmingly male and non–formally educated. Many of those identified in this study originate from villages within Maradi Department. Once they first leave the villages, migrants begin what eventually becomes a dual existence, which permits access to farmland during the rainy season as well as the opportunity to earn cash in the city during the off-season. Migrants often choose the closest viable destination: large enough to permit access to opportunities, but close enough to the village of origin to keep in touch. Of the almost 80 percent of my interviewees who were married, nearly half were accompanied by their families during their most recent trip to the city. On this same trip to the city, about half of the migrants traveled on foot.

An understanding of the intersection between individual circumstances and their larger environmental context is critical for pinpointing the motivations for temporary circular migration. The seasonality of precipitation plays a significant role. After the millet and sorghum harvest in October, migrants travel to the city to seek or resume urban employment for eight to ten months. Once the rains begin again in late May and early June, most migrants return to the village for two to four months. Migrants' responses to questions reveal that drought also plays a trigger role in affecting movements. It was commonly during drought years that migrants initiated a pattern of seasonal migration: 48 percent of the migrants interviewed left for the first time during or around the time of a drought. The first migrations of almost 44 percent of the migrants were motivated by household food shortages.

Through the activity of seasonal migration, which is essentially part-time residence in the city, village life is transplanted to the urban environment. After arriving, migrants pick up where they left off the preceding year, seeking or continuing employment with the help of social ties. Once in place in the city, migrants perform functions in the heart of the urban economy. My investigation identified the following broad categories of occupations: *post-harvest* activities, which usually involve some kind of biomass processing, including making fences and weaving palms; the *guarding* of wealthy houses in the city; a plethora of *services* and *crafts/semiskilled* occupations, which are traditional urban occupations; *women*'s occupations, including the preparation of foods, which are usually sold at the roadside, and prostitution; *market/sales* occupations involving the biweekly Maradi market or itinerant sales; *farming* in nearby fields or the seasonal river valley; *community leaders*, often *marabouts* (Islamic priests), who usually play the role of helping newer migrants become adjusted to the city; and the *nonemployed*, who are usually disabled.

On the average, migrants earn 763 FCFA (approximately US$1.50) per day. The vast majority work in the informal economy. Almost 58 percent satisfy only their basic needs in the city. Some, like guards, get pulled into year-round employment, gradually reducing the time they spend in the village, to the point where the only visits are infrequent social ones. The roles played by contacts, social networks, and experience in capitalizing on opportunities while maintaining ties to one's village are substantial. Almost 96 percent of migrants interviewed, even those who have moved to Maradi more or less permanently, maintain contact with their village in some way.

Widespread informal assistance is provided with the help of *marabouts* and commercial ties, creating a transplanted configuration of villages in the city. The result is that migrants are in effect displaced villagers, enjoying comparable housing and social outlets compared to friends and family back in the village but with the additional advantage of access to an urban economy. Despite lacking for little in the creature comforts available in the city, it is noteworthy that many stated a preference for remaining back in the village year-round.

The mobility response has been affected in recent years by changes in settlement and population dynamics. Many recently arriving migrants, particularly those doing post-harvest activities, come from small, later-settled villages in northern locations. Household migration decisionmaking is based on the perception that nonessential family members are obligated to leave the village periodically to save the family food and possibly earn cash that can be used to buy additional food. It is clear that the closing of the frontier and the resultant shortages of fertile land play a role in these decisions, though this role is indirect.

Certain basic facts about rural existence in the Sahel cannot be dismissed. Growing village populations cannot be supported on low and unreliable harvests, and from the analysis of intercensal population growth, it appears that populations are not growing in the northernmost and least-accessible villages. Declining yields with mounting risk of crop failure means that labor is better invested elsewhere. With land resources diminishing, the transition to nonagricultural livelihoods is assured. Concurrently, the importance of proximity to markets and outside opportunities is expanding, though the effect of this expansion on migration motivated by a need for cash is not necessarily distinct from migration that is motivated by food shortages. The new dependence on cash, mainly for food purchases and tax payments, means that seasonal migration becomes an increasingly rational response to changing circumstances.

To identify precisely how mobility is used as an economic strategy in the migrating population, I undertook an analysis of migrants' livelihoods. Chapter 6 examined the aforementioned occupational categories to answer questions about agricultural investments, remittances, and social ties. The livelihoods found in the city of Maradi had a wide range of portability. Some occupations required the presence of a market, some the presence of a monied class, some the presence of raw materials (such as palms for weaving), and some the presence of a nearby border. All the occupations required some manner of urban function. The livelihoods had their own particular histories, from the oldest and most portable crafts and service occupations to the more recently devised guard and post-harvest occupations. All of these show clear parallels between livelihoods that are chosen and changes that can be observed on the regional level.

Regardless of their livelihood, migrants maintain ties with their villages, usually to maintain food security by working their fields. More than 65 percent overall returned during the rainy season to farm, but only about 26 percent of those returning to farm said they had taken chemical fertilizer, seeds, or other inputs home with them on their last visit. The occupational categories differed, however, in how much responsibility their members had for farming. Those in crafts/semiskilled and service occupations frequently reported that they only helped out the family with farming back in the village, whereas over 70 percent of those in post-harvest occupations said that they themselves did the farming. One can conclude from this that the more recently evolving occupational groups are more essential to the village. And conversely, older migrant livelihoods are more detachable.

The remittances of money and goods back to the village are significant, though rarely sizable considering the level of earnings in the city. In all, almost 53 percent of the migrants interviewed replied that they sent items back to the village. Most remittances are annual lump sums of 1,000 to 5,000 to 10,000 FCFA (US$2–20), often made in the cold season in December and January after the migrant has had a chance to work for a few months in the city. Of these remittances, 93.5 percent were hand-delivered, either by the respondent or by friends, and 37 percent were intended for the purchase of food and other necessities to stop gaps in household supply. Viewed by occupational category, the market/sales and service occupations had the highest levels of remittances, as well as the highest incomes. The lowest levels of remittances were found in the post-harvest occupations.

Visits received in the city, letters, chance meetings at the market, and other social network mechanisms serve to link the migrant with the rural village. Over 75 percent of the migrants received visits from family members in the village. Social conventions govern the circulation of people in and out of Maradi. When visitors arrive, they are usually provided with gifts of food, pocket money, and clothing. Often their travel expenses, including the money for their return journey to the village, are paid for as well. The study found very low membership in formal organizations, but the critical importance of family and friends for migrants should not be discounted; they make the transition to the city not only possible but also enjoyable. Viewed by occupational category: 76.5 percent of those in post-harvest occupations received visits from family, as did 66.7 percent of the nonemployed; at the other end of the scale, only 20 to 25 percent of community leaders, craftspeople, and women received visits from family. The explanation for the low level of contact for women is that many urban women are divorced and have been shunned by their families. Some become prostitutes in the city, and many others practice more accepted traditional female occupations, such as food preparation or millet pounding.

For Chapter 7, the study returned to the village to explore four responses by rural communities to seasonal migration, revealing the effects of a rapidly urbanizing rural culture. Land-use change was the first of the four responses. Villagers and informants in Guidan Wari shared a common perception that agricultural fields are being robbed of their soil nutrients and that fertilizer is necessary. The second response was to change one's livelihood, which was regarded as a principal way for rural people to use economic diversification to adapt to changing circumstances. An increased use of social networks was the third response; this increase in social networks is marked by increasing levels of interaction—and control—of the urban realm over village life. Demographic change made up the fourth response, in that mobility acts to reduce rates of natural increase by moving village members elsewhere. In their social structure, the villages of Guidan Wari are similar to many rural peripheries throughout the world. They are populated by old people and abandoned spouses and children, with few young adult men. There is a palpable sense of severely limited opportunities. Those remaining in the village expressed feelings of a loss of cohesion and conviviality and a weakening of rural solidarity.

Community-level perceptions of seasonal migration's social consequences reveal profound social differences in views of seasonal migration between young and old and between men and women. The young tend to see mobility as inspiring, rewarding, or inevitable, whereas the old see mobility as a cultural as well as a political threat. Along gender lines, men view seasonal mobility as an opportunity to escape the drudgery of the "dead season" and to make better use of idle time, whereas women who are rooted in the village feel that they suffer the costs of male seasonal absence in the reduced number of economic options, even if they have fewer mouths to feed. Still, there is a diversity of responses that transcends these stereotypes. Though some women bemoaned the unfairness of being abandoned, others were indifferent to the male absence and were candid in stating they preferred not having their husbands around. In regard to gender relations, the irony of absence is noteworthy: Whether men are in the village or not makes no difference to some women because men are not in fact an integral part of daily household life even when present. Men often do not eat with their families, preferring the company of friends in the village, and they often interpret their supporting roles narrowly to minimize their obligation to their wives and children.

As I have demonstrated, circular mobility has clear social dimensions. Community responses also reveal certain environmental consequences of seasonal migration. This study's findings on the effects of circular migration on agriculture and natural resource management practices echoed the findings of existing research. For instance, in four case studies in the Sahel, male circular mobility did not itself induce environmentally degrading activities or increase pressure on renewable natural resources (David 1995). In my study of Guidan Wari, the impediment to village agricultural investment was the lack of cash. The desire for fertilizer, oxcarts, and other inputs and equipment is high, but they are not perceived to be affordable. This has put the villagers in a squeeze: They see the need to replace the soil nutrients that they have mined, but they have few options when they are so close to the hunger line. This fertilizer gap creates a vicious circle, impelling more and more farmers to leave the village to seek their fortunes elsewhere; and the labor supply available to undertake intensification efforts necessarily deteriorates. This connection between the nutrient squeeze and the resort to seasonal mobility was patently clear, and it was confirmed by villagers and migrants alike.

The analysis of migrants' activities and the effects of those actions on the sending villages shows that remittances seldom result directly in improvements. In all but a few more favored locations, where environmental conditions or proximity to markets make improvements economically feasible, seasonal or circular migration in Maradi Department is not the engine for rural development that agricultural economists and development practitioners would like it to be. There is very little evidence to show that migration earnings are fueling any sort of pervasive agricultural change in the villages of the Sahelian zone. The lack of capital and the high perceived level of risk were confirmed many times over. Therefore, the hypothesis of resource circularity is valid only if resources are defined very broadly.

But though migration does little to increase the food security of rural villages in the aggregate, it does increase the security of individual members by exposing them to more and more varied contacts. The rationale for movement is therefore more appropriately seen as network-based. Mobility may be draining remote villages of their labor, but this ultimately may not be as significant as the fact that nearly all migrants still maintain some form of contact with the village. This makes individuals, families, and communities less vulnerable to famine by spreading around risk, and having a diversity of locations increases options. This ultimately can be interpreted as the rationale for the movements observed.

My study of reorganization along the environmental margin in Niger, in "the periphery of a periphery," reveals the extent to which economy and environment are intertwined. It also reveals the geography of social and environmental change, to which heightened levels of circular migration are only one response. Given the community consequences of rapid urbanization, of which circular migration is a part, is village life a relic of Sahelian existence? Is circular migration a transitional mode that will eventually exclude the village? There are two practical answers to these questions. The rationale for maintaining contact with the village is based on the enduring need for food security. Even if they now live in the city, peasants still need the land where they can grow food. Even after they have long since moved away, villagers continue to maintain ties to their fathers' or grandfathers' villages out of familial obligation or because it is otherwise in their interest to keep social ties alive. Because contacts with villages require the continual rebuilding of social ties, *as long as migrants continue to be linked to their villages of origin*, the practice of circular mobility will continue to be stable. Circular migration is therefore transitional only so far as settlement of the Sahelian margin itself is transitional.

Circular migration should not be seen as a threat to the viability of the rural sector, I contend, but as the hope for the rural sector. As locational investment decisions, migrants' actions are significant. They are investing in their villages, but they are not investing money as much as time, building and maintaining external contacts that spread across space. Mobility therefore improves the viability of the rural sector. Mobility should be recognized for its ability to mitigate the effects of drought, to provide a steady year-round supply of cash, and to supersede the consequences of converting a marginal environment to permanent agriculture. As the village loses its function as the demographic engine, affective obligations shift over time to the city. And as succeeding generations gradually adopt a more urban outlook, the process of urbanization will gradually shift interests to the city. Cities are and always have been features of the Hausa landscape, and this transition has been underway for decades. It will not be finished for decades more.

Given the implications of a future in which a fifth of humanity is projected to be African,[3] the largest question raised by this study is whether the important social values of the village, such as the charity shown for hungry strangers or the ability to band together against famine, can be successfully transplanted to the city. By 2020 West Africa is estimated to have a population of 430 million (OECD 1994). The prospect of a seamless West African megalopolis extending from Abidjan to Douala, which cannot burst at the seams because it has no seams, appears almost mythical. It certainly seems very frightening to many. Questions arise about who will police this megalopolis, about who will maintain its public water systems. The inevitability of a massive population redistribution in West Africa away from the marginal fringe and toward coastal and inland cities has been recognized, yet moving out of the West African Sahel immediately is not an option for the approximately 50 million people who live there. If the benefits of cities could be transplanted into the villages of the environmental margin, then increased resources and infrastructure would enable rural villages to adapt to their changing circumstances. The process can be humane if allowed to proceed gradually, and if migrants are allowed to preserve their dignity.

The Inevitability of Change

This study of Maradi's *masu cin rani* took place more than twenty years after a devastating drought occurred in the Sahel. Prior to the drought, the region of the Sahel was all but unknown to the Western world. When the multiyear drought there captured the world's attention, an entity called "the Sahel" was created, seemingly out of whole cloth. After a tour of the drought-stricken region of the Sahel in 1974, then-UN Secretary-General

Kurt Waldheim was quoted as saying that unless aid is brought to the region fast, the countries of the Sahel region "could literally disappear" ("Drought for Democracy," *Time*, April 29, 1974, p. 42.) The great Sahelian famine was an inspiration in the 1970s for a subgenre of self-serving commentary about "lifeboat ethics" (Lucas and Ogletree 1976) that asked whether starving people in overpopulated lands were deserving of our help or not. "The Sahel has virtually nothing to offer the rest of the world," wrote Garrett Hardin (1977, p. 91), turning the region into an excuse for inaction. Evidence of suffering and the pity it inspires have always been an effective mechanism for continuing development assistance and increasing donations to humanitarian organizations. Whether aid comes or not, the swath of land lying between desert and savanna in West Africa and the people who inhabit or walk across it are not likely to disappear soon.

Because they place a low valuation on the specific site of settlement and a high valuation on networks and social ties, individuals and groups will decide to allocate resources to areas where the return is better or the risk of failure is lower. If the investors are mobile, there are reasons to expect the risk and the costs to be low. Starting up in a new location is relatively easy, given the social structure already in place. If the village is the community of individuals, *regardless of location*, then migrants are never truly "displaced." If places—like livelihoods—are portable, then migrants are trading little if anything away in their pursuit of opportunity and food security. The result of the dramatic increases in seasonal mobility has *disintegrated* village society but not *dissolved* it. Components have been removed and relocated elsewhere, person by person. Though parts of the Maradi region might be classified as vulnerable, by no means should the people moving about the region necessarily be considered so. Migrants are a product of social change, and they are capitalizing on their circumstances; only rarely are they victims of their circumstances.

In recounting depredations old and new in comparative perspective, it helps to take the long view. People who have over their history withstood the ravages of slave raiding, epidemics, and attacks by predators should perhaps be insulted that "we" think a half-century of inept European colonial occupation could destroy the fabric of society and stop civilization in its tracks. Where is the evidence that today's ravages are presenting a challenge against which the people are not equipped? The secret of the *masu cin rani* is that "they" are far better at surviving the challenges of the twenty-first century than "we" in the West believe them to be. Who survives, however—often the oldest and most influential as opposed to the children—might indeed be an ethical point of contention in the hypothetical debate between Sahelians and the West.

By extending this observation to activities to promote economic development, it is clear that policies to root peasants and to force them to invest in places instead of people are not in the peasants' interests. Development theory is moving beyond theories of staged transitions and toward an understanding of interactions and interrelatedness. The degree to which economy and ecology are mutually constituted in the Third World, where a sizable part of the population still derives its livelihood and security from managing or processing biomass, presents serious implications for how people perceive the places where they live.

The implication of a value system that defines *places* as constellations of people and commitments, instead of as inventories of physical contents, is that there is no apparent advantage to investing in villages to make them more permanent, unless such investments are perceived to be in the interests of the inhabitants. Instead of trying to deny or negate complicated social webs, concerned parties should acknowledge them and then work with them to make them more resilient. Community and livelihoods should be emphasized over permanence of settlement. If moving people establish priorities that transcend fixed locational coordinates, then food-for-work programs designed to keep people in their villages are nothing more than a new postcolonial form of forced labor to coerce people into performing tasks that would be perceived as undesirable or irrational without government subsidies or outside intervention. And efforts to "stanch the flow of migrants"—or to "control population," for that matter—without addressing underlying issues of land and labor needs are bound to fail.

As judged by its recent embrace of anthropology and geography, demography, that most quantitative of social sciences, is progressing beyond static numerical representations and toward more qualitative content. The importance of the regional context, especially in regard to driving forces that are not demographic, must still be realized. In the Sahelian zone, the relationship on the ground between environmental perceptions and demography is strong, as is the belief that people's fate is in the hands of divine powers. To the villagers, drought and lack of religious faith are seen as experientially linked, and the permanent loss of soil fertility, lack of rainfall, and famine are bound together with the abandonment of religious devotion. These perceptions are beliefs that should be respected. Conditions in the Sahelian zone, however, should not be allowed to deteriorate to the point where social mechanisms for surviving drought and famine violate the human rights of its inhabitants.

Rural culture should be given the opportunity to demonstrate that it is capable of dynamic change. There is a serious need for a unified approach by development organizations to emphasize the startling lack of control by Sahelian women over decisions regarding population and environment. In areas where circular migration is high, development programs must integrate health and natural resource management programs on the village level. Resource tenure systems must be changed to recognize formally women's power in the village, and this power should extend to inheritance laws; to ownership of household assets, fields, and livestock; to child custody; and to divorce. Women in Niger have mobilized since 1993 to gain more official rights, but this activity needs more than an endorsement by the national government; it needs to be implemented through the system of *arrondissements,* cantons, and villages. Doing so, however, would put the government directly at odds with elements of the Muslim establishment.

In the area of health and population, nongovernmental organizations need more resources to train mobile health extension workers to take affordable health care and family planning to the smallest and most distant villages. Quality improvement in the service at government medical centers must continue. Combined child and infant mortality must be reduced to below 200 per 1,000. When mortality is high, there is an almost insurmountable need to produce many children to overcome the effect of drought and survive for another generation. A mortality rate of 318 per 1,000 for infants and children under five (DHS 1992) is still far too high to permit any reduction in fertility. Village meetings to discuss local public health and family planning must receive the endorsement of government and religious authorities.

In the area of market and transport planning, development banks should target infrastructure improvement toward making lower-order market roads more accessible. Circular labor migration should be encouraged, and owner-investment in passenger vehicles should be made easier. The free flow of people in the Sahel should be made a national priority. Migrants' associations should be revived and officially sanctioned, bureaucratic obstacles to free movement should be reduced, and agreements with other West African states pertaining to labor migration should be strengthened. These will have consequences, such as increased labor productivity and deeper markets for services and goods, that will ultimately reduce the vulnerability to drought-induced famine. Until this happens, the alphabet soup of acronyms offered by the development community will not make a difference.

This research suggests a number of implications for the discipline of geography. First, the idea that causes and consequences of patterns and variations in the living world can be different in different places is particularly underutilized. The discipline's power to depict such phenomena panoramically is the geographer's calling card and constitutes a unique contribution to science and policy. In my study, the example of vulnerability is instructive. I identified patterns of circular mobility that are distinctive to West Africa and posed questions about migrants' abilities to use movement to mitigate the effects of drought and chronic food shortages. How can circular mobility simultaneously contribute to *and* alleviate social and biophysical vulnerability? Why do people migrate away from the villages where their food is grown and toward cities to earn cash, which is then spent on food? Geographers have the tools to explain how mobility patterns are both temporally dynamic and spatially situated, and such explanations can shed light on what social scientists label "hidden heterogeneity." Situational commonalities such as environmental and economic marginality, which transcend the spatial primitives of latitude and longitude, affirm the existence of community-based configurations. The nexus of gender inequality in decisionmaking, a stressed natural resource base, and the high value of labor in biomass-based subsistence economies (Dasgupta 1993) is self-reinforcing and will be a continued focus of concern throughout the next century.

Second, it should once and for all be acknowledged that "the environment," particularly at the scale of human lives, has an undeniable effect on behavior and needs to be reexamined using geographers' tools. I have suggested that climate in Sahelian West Africa—specifically intra-annual and interannual precipitation variability—plays a role that cannot be blamed on colonialism or explained away by social constructionists. My study found that population mobility was heavily influenced by precipitation rather than exclusively determined by it. People's movements in the region have a distinctive circularity to them that is eminently rational for balancing needs and obligations amid an environment of risk and uncertainty. In his edited volume on seasonality, Robert Chambers writes:

> Climate has been out of fashion as an explanation of poverty; but the location of richer countries in temperate latitudes and poorer countries in the tropics is so marked that climatic factors cannot lightly be dismissed, whether they are fashionable or not. One possibility is that climatic influences have been underestimated because of a failure to see that seasonally adverse factors interact and reinforce each other at certain times of the year. Tropical seasonality tends to be overlooked anyway; and where it is noticed, it is usually along a single disciplinary dimension. (Chambers, Longhurst, and Pacey 1981, p. 1)

The idea that the environment influences behavior was advanced early in this century and was later renounced for being misguided pseudoscience. But the fact that environmental determinism was criticized owed more to the overt and misguided racism of the day than to its scientific grounding. In continuing to harbor guilt over its historical championing of environmental determinism, geographers have been playing hurt for too long. The legacy of guilt has continued to the point where geographers still find themselves wary of vestigial prejudices that have nothing to do with the reality on the ground that they are trying to comprehend and depict. Current interests in population and environment should be based on convictions that are both humanistic and "green." Instead of denying the perceived intrusions of the natural world, we should welcome them into our daily actions and into our culture.

Third, I have suggested the broad outlines for a proposed "grounded population geography," which deserves to be embraced and should be broadened to include a greater emphasis on concepts from human geography, including place, culture, and environment. In regard to human reproduction and cultural practices, population geographers and demographers should be aware that place *matters*. Behavior is situated. How a region was settled, the social systems set in place, patterns of relations, expressions of social power, and forums for managing conflicts—all reinforce the influence of local practices. The chronology and conditions under which a settlement was formed and the extent to which it is self-contained or driven by outside interactions within a regional system provide the grounds for comparisons with other places. All these make the influence of places impossible to hold constant in a social science balancing equation. Culture is a series of accumulated actions over time, situated in a particular place. Local analysis of reproduction, birth and death, livelihood, and movement must go beyond simple descriptions toward the ways that perceptions and patterns of behavior are interconnected. As for environment, this study provided a basis for understanding one way that people achieve food sufficiency, and this is situated in a local environment of particular soils, rainfall, and topography. The strategies used by individuals, households, and communities to gain food security all stem from features of both the local environment and the system that links the local to the global. Causes and consequences of the geography of population will be different in different places. Geographers must swallow the squeamishness brought on by polemical population debates and realize that population is not solely numbers, that it is also people. Population is not limited to associations with food and natural resources; rather, it also encompasses the human image, landscape, and spirit.

Fourth, geography as a discipline must transcend the global view and embrace the local by including individual perspectives, identifying in particular the power of people's voices and their stories. The field should not be deprived of the organic stirrings from "below" that have been witnessed in the humanities. A new populist geography, which is invigorated by contestation and coalition building even as it is scientifically grounded, should affirm universal human qualities; it should not, however, turn people into poster children for whichever critical theory is in vogue. To illuminate everyday lives, a populist geography must present people and places and events in a nonreductive, nonjudgmental way, using personal testimonies, life histories, maps of movements in time and space, settlement histories, and perceptions of the living world, which *necessarily* consists of both human and physical realms. Geographers need to derive inspiration and content *from* the world; they also need to speak *to* the world, which includes those outside the discipline. There is a huge unrealized potential for doing what social scientists have long discouraged: We need to identify with our subjects.

The task of understanding the changing worlds of circular migrants in Niger is based on the commonsense observation that much of the growth occurring in the next century will be in the places least able to absorb the increases. Given the likelihood that 10 billion people will be walking the planet in the next century, the ways that poor people live and adapt to change must be understood. Current frameworks of social and environmental change do not accurately explain or reflect the reality of the 1.3 billion people who earn less than a dollar a day, receive inadequate nutrition, and drink unclean water and who are labeled with the name "poor" (UNDP 1997). The Third World has a social structure, resilience, and respect for affective ties that serve as examples to all people. These attributes should not be threatened, either by Western-style development or through shortsighted or ineffectual policies. As the analysis showed, migrants are neither desperate harbingers of future disorder nor saviors of economic development programs. By moving, migrants merely act in their own interests, step by step. We must look beyond the chaotic street scenes of Third World cities to see the real people who know how to adapt and survive. If we do not, we will be blind to the underlying and as yet unfulfilled potential of collective action.

Notes

1. This might explain why migrants in Maradi tend to remit food and not fertilizer to their families in the village.

2. There are indications that Niger's dependence on countries to its south may be strained further. As the following news dispatch relates, coastal migration destinations for Nigériens are facing popular opposition:

"At Least One Nigérien Murdered in Côte d'Ivoire Anti-Hausa Riots

"A wave of anti-Hausa hysteria in Côte d'Ivoire last weekend led to mob violence during which six Hausa, at least one of whom was Nigérien, were murdered.

"The killings were sparked by rumors that have been circulating in Côte d'Ivoire for several weeks. Some Ivoiriens claimed that Hausa people were using magic to make men's penises and women's breasts disappear, and then extorting money from their victims before restoring the missing organs. Last weekend, a Hausa man from Niger was burned to death in Abidjan by a mob that accused him of causing a teenager's penis to vanish. The hysteria spread throughout the city, with thousands of youths looting shops and attacking the homes of foreigners" (*Johannesburg Mail and Guardian,* Agence France Presse, Voice of America, March 1997).

3. The continent of Africa contained 8.9 percent of the total world population in 1950. Extrapolating from current rates of population growth, it will have 20.1 percent in 2025 (McNicoll 1984). This assumes that the catastrophic progression of HIV through the population is slowed. Of the estimated 15 million cases of HIV worldwide, about 9 million are in Africa, though there are disagreements about the accuracy of the estimate.

Bibliography

Abadie, Maurice. 1927. *La Colonie du Niger*. Paris: Société d'Editions Géographiques, Maritimes et Coloniales.

Adejuwon, J. O., E. E. Balogun, and S. A. Adejuwon. 1990. On the annual and seasonal patterns of rainfall fluctuations in sub-Saharan West Africa. *International Journal of Climatology* 10: 839–848.

Adepoju, Aderanti. 1974. Migration and socio-economic links between urban migrants and their home communities in Nigeria. *Africa* 44, 4: 383–396.

_____. 1981. Issues in the study of migration and urbanization in Africa. In *Population Movements: Their Forms and Functions in Urbanization and Development*, ed. P. A. Morrison. Liège, Belgium: IUSSP.

_____. 1994. Preliminary analysis of emigration dynamics in sub-Saharan Africa. *International Migration* 32, 2: 197–216.

Agnew, Clive. 1990. Spatial aspects of drought in the Sahel. *Journal of Arid Environments* 18: 279–293.

Allan, William. 1949. *Studies in African Land Usage in Northern Rhodesia*. Rhodes-Livingstone Papers, no. 15. London: Oxford University Press.

_____. 1965. *The African Husbandman*. New York: Barnes and Noble.

Amin, Samir, ed. 1974a. *Modern Migrations in Western Africa*. London: Oxford University Press.

_____. 1974b. *Neo-Colonialism in West Africa*. New York: Monthly Review Press.

Ancey, G. 1988. Etude du secteur agricole du Niger: Les politiques. Rapport de mission, SEDES, Paris.

Anker, Richard. 1989. Measuring women's participation in the African labour force. In *The Informal Economy*, ed. A. Portes, M. Castells, and L. A. Benton. Baltimore: Johns Hopkins University Press.

Aubreville, André. 1949. *Climat, Forets, et Désertification de l'Afrique Tropicale*. Paris: Société d'Editions Géographiques, Maritimes et Coloniales.

Axinn, William G., Thomas E. Fricke, and Arland Thornton. 1991. The microdemographic community-study approach: Improving survey data by integrating the ethnographic method. *Sociological Methods and Research* 20, 2: 187–217.

Baier, Stephen. 1980. *An Economic History of Central Niger*. Oxford: Clarendon Press.

Barou, Jacques. 1976. L'émigration dans un village du Niger. *Cahiers d'Etudes Africaines* 16, 3: 627–632.

Barth, Heinrich. 1857. *Travels and Discoveries in North and Central Africa*. New York: Harper and Brothers.

Becker, Charles, Andrew Hamer, and Andrew Morrison. 1994. *Beyond Urban Bias in Africa*. Portsmouth, NH: Heinemann.

Berg, Elliot J. 1965. The economics of the migrant labor system. In *Urbanization and Migration in West Africa*, ed. H. Kuper. Berkeley: University of California Press.

Berry, Sara. 1984. The food crisis and agrarian change in Africa: A review essay. *African Studies Review* 27, 2: 59–112.

_____. 1989. Social institutions and access to resources in African agriculture. *Africa* 59, 1: 41–55.

_____. 1993. *No Condition Is Permanent*. Madison: University of Wisconsin Press.

Bilsborrow, Richard E. 1981. Surveys of internal migration in low-income countries: The need for and content of community-level variables. Population and Labour Policies Programme, Working Paper no. 98. Geneva: International Labour Office.

Bilsborrow, Richard E., and Hania Zlotnik. 1995. The systems approach and the measurement of the determinants of international migration. In *International Migration: Issues in Measurement and Analysis*, ed. Rob van der Erf and Lisbeth Heering. Luxembourg: Eurostat.

Binswanger, Hans P., and Prabhu L. Pingali. 1988. Technological priorities for farming in sub-Saharan Africa. *World Bank Research Observer* 3, 1: 81–98.

Blaikie, Piers. 1989. Environment and access to resources in Africa. *Africa* 59, 1: 18–40.

Blaikie, Piers, and Harold Brookfield. 1987. *Land Degradation and Society*. London: Methuen.

Blaikie, Piers, Terry Cannon, Ian David, and Ben Wisner. 1994. *At Risk: Natural Hazards, People's Vulnerability, and Disasters*. New York: Routledge.

Bohannan, P., and G. Dalton, eds. 1962. *Markets in Africa*. Evanston, IL: Northwestern University Press.

Boserup, Ester. 1965. *The Conditions of Agricultural Growth*. Chicago: Aldine Publishing Co.

_____. 1970. *Women's Role in Economic Development*. London: George Allen and Unwin.

_____. 1981. *Population and Technological Change*. Chicago: University of Chicago Press.

_____. 1990. Population, the status of women, and rural development. In *Rural Development and Population: Institutions and Policy*, supplement to volume 15, *Population and Development Review*, ed. G. McNicoll and M. Cain. New York: Oxford University Press.

Boye, Abd-el Kader, Kathleen Hill, Stephen Isaacs, and Deborah Gordis. 1991. Marriage law and practice in the Sahel. *Studies in Family Planning* 22, 6: 343–349.

Brasset, P., J. Koechlin, and C. Raynaut. 1984. *Rapport de Mission Socio-Géographique: Proposition pour un Zonage Agro-Ecologique du Département de Maradi*. Bordeaux: Université de Bordeaux II.

Brunner, Jake, Kevin Dalsted, and Ari Arimi. 1995. The use of aerial video for land use/land cover characterization and natural resource management project impact assessment in Niger. Paper prepared for USAID/Niger.

Bryson, Reid. 1973. Drought in Sahelia. *The Ecologist* 3: 366–371.

Burton, Ian, Robert W. Kates, and Gilbert White. 1978. *The Environment as Hazard.* New York: Oxford University Press.

Butzer, Karl. 1983. Late quaternary environmental change in the Sahel. In *Environmental Change in the West African Sahel.* Advisory Committee on the Sahel. Washington, DC: National Academy Press.

Cain, M. 1978. The household lifecycle and economic mobility in Bangladesh. Centre for Policy Studies, Working Paper. New York: The Population Council.

Cain, M., and Geoffrey McNicoll. 1988. Population growth and agrarian outcomes. In *Population, Food, and Rural Development,* ed. R. D. Lee et al. Oxford: Clarendon Press.

Caldwell, John C. 1969. *Africa Rural-Urban Migration: The Movement to Ghana's Towns.* Canberra: Australian National University Press.

_____. 1975. The Sahelian drought and its demographic implications.Report, Bamako, Mali. Washington, DC: Overseas Liaison Committee.

Caldwell, John C., Allan G. Hill, and Valerie J. Hull. 1988. *Micro-Approaches to Demographic Research.* London: Kegan Paul International.

Carter, Michael, and Frederic Zimmerman. 1993. Risk, scarcity, and land market: The uneven economics of induced institutional change in the West African Sahel. Working Paper no. 86. College Park, MD: Center for Institutional Reform and the Informal Sector.

Centre des Etudes et Recherches pour la Population et de Developpement. 1993. *Réseau Migrations et Urbanisation en Afrique de l'Ouest.* Bamako, Mali: CERPOD.

Chambers, Robert. 1983. *Rural Development: Putting the Last First.* Essex, England: Longman Scientific and Technical.

Chambers, Robert, Richard Longhurst, and Arnold Pacey, eds. 1981. *Seasonal Dimensions to Rural Poverty.* London: F. Pinter.

Chambers, Robert, Arnold Pacey, and Lori Ann Thrupp, eds. 1989. *Farmer First: Farmer Innovation and Agricultural Research.* New York: Bootstrap Press.

Chapman, Murray. 1975. Mobility in a non-literate society: Method and analysis for two Guadalcanal communities. In *People on the Move: Studies on Internal Migration,* ed. L. A. Kosinski and R. M. Prothero. London: Methuen.

_____. 1988. Population movement studied at micro-scale: Experience and extrapolation. In *Micro-approaches to Demographic Research,* ed. J. C. Caldwell, A. G. Hill, and V. J. Hull. London: Kegan Paul International.

Charlick, Robert. 1991. *Niger: Personal Rule and Survival in the Sahel.* Boulder, CO: Westview Press.

Charney, Jules. 1975. Dynamics of desert and drought in the Sahel. *Quarterly Journal of the Royal Meteorological Society* 101: 193–202.

Clapperton, Hugh. 1829. *Journal of a Second Expedition into the Interior of Africa from the Bight of Benin to Soccatoo.* London: John Murray.

Clark, William A. V. 1986. *Human Migration.* Beverly Hills, CA: Sage Publications.

Club du Sahel. 1995. *Club du Sahel Newsletter.* Paris: OECD.

Cohen, Abner. 1969. *Custom and Politics in Urban Africa: A Study of Hausa Migrants in Yoruba Towns.* Berkeley and Los Angeles: University of Calfornia Press.

Copans, Jean. 1983. The Sahelian drought: Social sciences and the political economy of underdevelopment. In *Interpretations of Calamity*, ed. Kenneth Hewitt.
Boston: Allen and Unwin.

Coquery-Vidrovitch, C. 1988. *Africa: Endurance and Change South of the Sahara*.
Berkeley and Los Angeles: University of California Press.

Coquery-Vidrovitch, C., and Paul Lovejoy, eds. 1985. *The Workers of African Trade*.
Beverly Hills, CA: Sage Publications.

Cordell, D., J. Gregory, and V. Piché, eds. 1996. *Hoe and Wage*. Boulder, CO: Westview Press.

Cour, Jean-Marie. 1994. Population trends and economic growth in sub-Saharan
Africa. In *Involuntary Resettlement in Africa*, ed. C. C. Cook. World Bank Tech-
nical Paper no. 227. Africa Technical Department Series. Washington DC:
World Bank.

Cross, Nigel, and Rhiannon Barker, eds. 1991. *At the Desert's Edge: Oral Histories
from the Sahel*. London: Panos Publications.

Curry, John. 1989. Occupation and drought vulnerability: Case studies from a village in Niger. In *African Food Systems in Crisis*, part 1: *Micro-perspectives*, ed. R.
Huss-Ashmore and S. Katz. New York: Gordon and Breach.

Dankoussou, I., S. Diarra, D. Laya, and D. A. Pool. 1975. Niger. In *Population
Growth and Socio-Economic Change in West Africa*, ed. J. C. Caldwell. New York:
Columbia University Press.

Dasgupta, Partha. 1993. *An Inquiry into Well-Being and Destitution*. Oxford:
Clarendon Press.

Dasgupta, Partha, and Karl-Göran Mäler. 1994. Poverty, institutions, and the environmental resource base. World Bank Environment Paper no. 9. Washington
DC: World Bank.

David, Philippe. 1969. Maradi précolonial: L'état et la ville (République du
Niger). *Bulletin de l'Institut Français d'Afrique Noire*, ser. B, 31, 3: 638–688.

David, Rosalind. 1995. *Changing Places? Women, Resource Management, and Migration in the Sahel*. London: SOS Sahel.

Davidson, Basil. 1965. *A History of West Africa Prior to the Nineteenth Century*. London: Longmans, Green.

De Waal, Alexander. 1988. Famine early warning systems and the use of socio-
economic data. *Disasters* 12, 1: 81–91.

———. 1989. *Famine That Kills*. Oxford: Clarendon Press.

Delehanty, James M. 1988. The Northward Expansion of the Farming Frontier on
Twentieth Century Central Niger. Ph.D. diss., University of Minnesota Department of Geography.

Demko, George J., and William B. Wood, eds. 1994. *Reordering the World: Geopolitical Perspectives of the Twenty-first Century*. Boulder, CO: Westview.

DHS (Demographic and Health Surveys). 1992. *Enquête Démographique et de Santé,
Niger*. Columbia, MD: Macro International.

Direction de l'Agriculture, Département de Maradi, République du Niger. 1996.
Rapport Annuel des Activités d'Hivernage, Département de Maradi. Rapports,
1988–1996. Niamey, Niger: Government of Niger.

Dos Santos, Theotonio. 1971. The structure of dependence. *American Economics
Review* 60, 2: 231–236.

Downing, Thomas E. 1990. Monitoring and responding to famine: Lessons from the 1984–85 food crisis in Kenya. *Disasters* 14, 3: 204–229.

Duffill, M. B., and P. Lovejoy. 1985. Merchants, porters, and teamsters in the nineteenth century central Sudan. In *The Workers of African Trade,* ed. C. Coquery-Vidrovitch and P. Lovejoy. Beverly Hills, CA: Sage Publications.

Dyson, Tim. 1996. *Population and Food: Global Trends and Future Prospects.* London: Routledge.

Ehrlich, Paul. 1968. *The Population Bomb.* New York: Ballantine.

El-Hinnawi, E. 1985. *Environmental Refugees.* Kenya: United Nations Environment Program.

Elbow, Kent Michael. 1992. Popular Participation in the Management of Natural Resources: Lessons from Baban Rafi, Niger. Ph.D. diss., University of Wisconsin–Madison.

Ezenwe, Uka. 1983. *ECOWAS and the Economic Integration of West Africa.* New York: St. Martin's Press.

Falloux, F., and A. Mukendi, eds. 1988. *Desertification Control and Renewable Resource Management in the Sahelian and Sudanian Zones of West Africa.* World Bank Technical Paper no. 70. Washington, DC: World Bank.

FAO (Food and Agriculture Organization). 1990. Programme Complet de Sécurité Alimentaire du Niger. Tableaux Multicritères par Arrondissement. FAO Project GCPS/NER/031/NOR.

———. 1995. *Production Yearbook.* Rome: FAO.

Faulkingham, R., and P. Thorbahn. 1975. Population dynamics and drought: A village in Niger. *Population Studies* 29, 3: 463–477.

FEWS (Famine Early Warning System). 1996. Vulnerability assessment. Washington: FEWS.

FEWS (Famine Early Warning System). 1997. Vulnerability assessment. Washington: FEWS.

Findlay, M. I. 1975. *The Ancient Economy.* London. Reprint, Berkeley and Los Angeles: University of California Press, 1999.

Findlay, Sally. 1987. *Rural Development and Migration.* Boulder, CO: Westview.

———. 1992. Circulation as a drought-coping strategy in rural Mali. In *Migration, Population Structure, and Redistribution Policies,* ed. C. Goldscheider. Boulder, CO: Westview Press.

Fleuret, A. 1986. Indigenous responses to drought in sub-Saharan Africa. *Disasters* 10, 3: 224–229.

Food and Agriculture Organization/International Institute for Applied Systems Analysis. 1982. *Potential Population Supporting Capacities of Lands in the Developing World.* Technical Report of Project FPA/INT 1513. Rome: FAO.

Foote, Karen, Kenneth Hill, and Linda Martin, eds. 1993. *Demographic Change in Sub-Saharan Africa.* Washington, D.C.: National Academy Press.

Frank, André Gunder. 1967. *Capitalism and Underdevelopment in Latin America.* New York: Monthly Review Press.

Franke, Richard, and Barbara Chasin. 1980. *Seeds of Famine.* Totowa, NJ: Allenheld and Osmun.

French, Howard. 1996. Migrant workers take AIDS risk home to Niger. *New York Times,* February 8, A3.

Freudenberger, Mark S., and Paul Mathieu. 1993. *The Question of the Commons in the Sahel*. Madison, WI: Land Tenure Center.

Fuglestad, Finn. 1983. *A History of Niger, 1850–1960*. Cambridge: Cambridge University Press.

Gado, Boureima Alpha. 1993. *Une Histoire des Famines au Sahel: Etude des Grandes Crises Alimentaires, XIXe-XXe Siècles*. Paris: Editions l'Harmattan.

Geertz, Clifford. 1963. *Agricultural Involution: The Process of Ecological Change in Indonesia*. Berkeley and Los Angeles: University of California Press.

Germani, Gino. 1965. Migration and acculturation. In *Handbook for Social Research in Urban Areas*, ed. P. M. Hauser. Paris: UNESCO.

Glantz, Michael H., ed. 1976. *The Politics of Natural Disaster: The Case of the Sahel Drought*. New York: Praeger.

_____, ed. 1986. *Drought and Hunger in Africa: Denying Famine a Future*. New York: Cambridge University Press.

Goldschmidt-Clermont, Luisella. 1994. Assessing women's economic contributions in domestic and related activities. In *Gender, Work, and Population in Sub-Saharan Africa* ed. A. Adepoju and C. Oppong. Portsmouth, NH: Heinemann.

Gould, W. T. S., and R. M. Prothero. 1975. Space and time in African population mobility. In *People on the Move: Studies on Internal Migration*, ed. L. Kosinski and R. M. Prothero. London: Methuen.

Gowers, W. F. 1911. Notes on Trade in Sokoto Province. Manuscript in Rhodes Collection, Oxford.

Grégoire, Immanuel. 1992. *The Alhazai of Maradi*. Boulder, CO: Lynne Rienner.

_____. 1997. Major Sahelian trade networks: Past and present. In *Societies and Nature in the Sahel*, ed. Claude Raynaut. New York: Routledge.

Grégoire, Immanuel, and Claude Raynaut. 1980. *Présentation Générale du Département du Maradi*. Bordeaux: Université de Bordeaux II.

GRID (Groupe de Recherches Interdisciplinaires [Claude Herry, Emmanuel Grégoire, and Anne Luxureau]). 1990. *Urbanisation et Santé à Maradi*. Bordeaux, France: Université de Bordeaux II.

Grove, A. T. 1961. Population and agriculture in northern Nigeria. In *Essays on African Population*, ed. K. M. Barbour and R. M. Prothero. London: Routledge and Kegan Paul.

Gugler, J., and W. G. Flanagan. 1978. Urban-rural ties in West Africa: Extent, interpretation, prospect, and implications. In *Migration and the Transformation of Modern African Society*, ed. W. M. J. Van Binsbergen and H. A. Meilink. African Perspectives 1978/1. Leiden, Netherlands: Africa-Studie-Centrum.

Haberkorn, Gerald. 1992. Temporary versus permanent population mobility in Melanesia: A case study from Vanuatu. *International Migration Review* 26, 3: 806–842.

Halfacree, Keith, and Paul Boyle. 1993. The challenge facing migration research: The case for a biographical approach. *Progress in Human Geography* 17, 3: 333–348.

Hardin, Garrett. 1974. Living on a lifeboat. *BioScience* 24, 10: 561–568.

_____. 1977. *The Limits of Altruism: An Ecologist's View of Survival*. Bloomington: Indiana University Press.

Harris, F. M. A., and B. W. Bache. 1995. Nutrient budgets in relation to the sustainability of indigenous farming systems in Northern Nigeria. Agronomy and Cropping Systems Programme, Natural Resources Institute.

Harrison, Paul. 1992. *The Third Revolution: Environment, Population, and a Sustainable World*. New York: St. Martin's Press.

_____. 1996. *Caring for the Future*. Produced by the Independent Commission on Population and Quality of Life. New York: Oxford University Press.

Haswell, M. R. 1953. *Economics of Agriculture in a Savannah Village*. London: Colonial Office, Her Majesty's Stationery Office.

Hewitt, K., ed. 1983. *Interpretations of Calamity*. Boston: Allen and Unwin.

Hill, Allan G. 1990. Demographic responses to food shortages in the Sahel. In *Rural Development and Population: Institutions and Policy*, supplement to volume 15, *Population and Development Review*, ed. G. McNicoll and M. Cain. New York: Oxford University Press.

Hill, Polly. 1972. *Rural Hausa*. Cambridge: Cambridge University Press.

Homer-Dixon, Thomas F. 1991. On the threshold: Environmental changes as causes of acute conflict. *International Security* 16, 2: 76–116.

Hopkins, A. G. 1973. *An Economic History of West Africa*. London: Longman.

Hopkins, Jane, and Thomas Reardon. 1992. Potential welfare impacts of trade regime changes on rural households in Niger: A focus on cross-border trade. Paper prepared by the IFPRI/ISRA Conference on the Regional Integration of Agricultural Markets in West Africa: Issues for Sahelian Countries.

_____. 1993. Agricultural price policy reform impacts and food aid targeting in Niger. IFPRI-INRAN Report (unpublished) to USAID Niamey.

Hopkins, Jane, and Ellen Taylor-Powell, eds. 1992. *Perceptions of Famine and Food Insecurity in Rural Niger*. USAID Working Papers, vol. 1. Washington, DC: USAID.

Hugo, Graeme J. 1981. Village-community ties, village norms, and ethnic and social networks: A review of evidence from the third world. In *Migration Decision Making*, ed. G. De Jong and R. W. Gardner. New York: Pergamon.

Hugon, Philippe. 1980. Les petites activités marchandes et l'emploi de secteur "informel": Le cas africain. 1977. Paris: Université de Paris I, Institute d'Etude de Développement Economique et Social, Groupes de Recherche.

Hulme, M., and M. Kelly. 1993. Exploring the links between desertification and climate change. *Environment* 35, 6: 5–45.

Huntington, Samuel. 1996. *The Clash of Civilizations and the Remaking of World Order*. New York: Simon and Schuster.

Hutchinson, C. F. 1991. Uses of satellite data for famine early warning in sub-Saharan Africa. *International Journal of Remote Sensing* 12, 6: 1405–1421.

Iliffe, John. 1987. *The African Poor*. Cambridge: Cambridge University Press.

Jacobson, Jody. 1988. Environmental refugees: A yardstick of habitability. Worldwatch Paper no. 86. Washington, DC: Worldwatch Institute.

Johnson, Willard R., and Vivian R. Johnson. 1990. *West African Governments and Volunteer Development Organizations*. Lanham, MD: University Press of America.

Jones, Emlyn S. 1995. Community analysis for Guidan Wari. Unpublished report for Peace Corps Niger. Niamey: Peace Corps Niger.

Jones, Huw. 1990. *Population Geography*. London: Paul Chapman.

Kaplan, Robert. 1994. The coming anarchy. *Atlantic Monthly* 273, 2 (February): 44–76.

———. 1996. *The Ends of the Earth: A Journey at the Dawn of the Twenty-first Century*. New York: Random House.

Kates, Robert, and Viola Haarmann. 1992. Where the poor live. *Environment* 34, 4: 5–28.

Kearney, Michael. 1986. From the invisible hand to visible feet: Anthropological studies of migration and development. *Annual Review of Anthropology* 15: 331–361.

Kelly, Charles. 1994. Bringing population dynamics into food balance sheet calculations. *Disasters* 18, 2: 171–176.

Kennedy, Paul. 1993. *Preparing for the Twenty-first Century*. New York: Random House.

Koechlin, J., C. Raynaut, and M. Stigliano. 1980. *Occupation Agricole en 1975 et Aptitudes du Milieu dans le Département du Maradi*. Bordeaux: Université de Bordeaux II.

Kopytoff, Igor. 1987. *The African Frontier*. Bloomington: Indiana University Press.

Kosinski, L., and R. M. Prothero, eds. 1975. *People on the Move: Studies on Internal Migration*. London: Methuen.

Lawson, Victoria, and Lynn Staeheli. 1990. Realism and the practice of geography. *Professional Geographer* 42, 1: 13–20.

Lee, Everett. 1969. A theory of migration. In *Migration*, ed. J. A. Jackson. Cambridge: Cambridge University Press.

Lee, R. D., W. B. Arthur, A. C. Kelley, G. Rodgers, and T. N. Srinivasan, eds. 1988. *Population, Food, and Rural Development*. Oxford: Clarendon Press.

Lele, Uma, and Steven W. Stone. 1989. Population pressure, the environment, and agricultural intensification. MADIA Discussion Paper no. 4. Washington, DC: World Bank.

Lesthaeghe, Ronald. 1986. On the adaptation of sub-Saharan systems of reproduction. In *The State of Population Theory*, ed. D. Colemanand R. Schofield. Oxford: Blackwell.

Lesthaeghe, Ronald, ed. 1989. *Reproduction and Social Organization in sub-Saharan Africa*. Berkeley and Los Angeles: University of California Press.

Liverman, Diana. 1989. Vulnerability to global environmental change. In *Understanding Global Environmental Change*, ed. Roger Kasperson, Kirsten Dow, Dominic Golding, and Jeanne Kasperson. Earth Transformed Program. Worcester, MA: Clark University.

Loofboro, Lynne. 1993. Relations in three agropastoral villages: A framework for analyzing natural resource use and environmental change in the arrondissement of Boboye, Niger. Discussion Paper no. 4, Land Tenure Center, University of Wisconsin, Madison.

Lovejoy, Paul E. 1980. *Caravans of Kola: The Hausa Kola Trade, 1700–1900*. Zaria and Ibadan: Ahmadu Bello University Press.

Lovejoy, Paul E., and Steven Baier. 1975. The desert-side economy of the Central Sudan. *International Journal of African Historical Studies* 7, 4: 551–581.

Lubeck, Paul. 1986. *Islam and Urban Labor in Northern Nigeria: The Making of a Muslim Working Class.* Cambridge: Cambridge University Press.

Lucas, George R. Jr., and Thomas W. Ogletree, eds. 1976. *Lifeboat Ethics: The Moral Dilemmas of World Hunger.* New York: Harper and Row.

Lugard, Frederick D. 1902. *Colonial Reports: Annual #476, Northern Nigeria Report for 1904.* London: His Majesty's Stationery Office.

Mabogunje, Akin. 1970. Systems approach to a theory of rural-urban migration. *Geographical Analysis* 2, 1: 1–18.

_____. 1972. *Regional Mobility and Resource Development in West Africa.* Montreal: McGill-Queen's University Press.

_____. 1990. Agrarian responses to outmigration in sub-Saharan Africa. In *Rural Development and Population: Institutions and Policy,* supplement to volume 15, *Population and Development Review,* ed. G. McNicoll and M. Cain. New York: Oxford University Press.

Malthus, Thomas R. 1817. *An Essay on the Principle of Population.* 5th ed. London: J. Murray.

Martin, Phyllis, and Patrick O'Meara, eds. 1986. *Africa.* 2nd ed. Bloomington: Indiana University Press.

Massey, Douglas. 1987. The ethnosurvey in theory and practice. *International Migration Review* 21: 1498–1522.

McGregor, Jo Ann. 1994. Climate change and involuntary migration: Implications for food security. *Food Policy* 19, 2: 120–132.

McNicoll, Geoffrey. 1984. Consequences of rapid population growth: An overview and assessment. *Population and Development Review* 10: 177–240.

_____. 1990. Institutional effects on rural economic and demographic change. In *Rural Development and Population: Institutions and Policy,* supplement to volume 15, *Population and Development Review,* ed. G. McNicoll and M. Cain. New York: Oxford University Press

McNicoll, Geoffrey, and Mead Cain. 1988. Population growth and agrarian outcomes. In *Population, Food, and Rural Development,* ed. R. D. Lee et al. Oxford: Clarendon Press.

McNicoll, Geoffrey, and Mead Cain, eds. 1990. *Rural Development and Population: Institutions and Policy,* supplement to volume 15, *Population and Development Review.* New York: Oxford University Press.

Meillassoux, Claude, ed. 1971. *The Development of Indigenous Trade and Markets in West Africa.* London: Oxford University Press.

_____. 1974. Development or exploitation: Is the Sahel famine good business? *Review of African Political Economy* 1: 27–33.

_____. 1981a. Paysans africains et travailleurs immigrés: De la surexploitation au génocide par la faim. Report by Centre Tricontinental, Paris. Louvain-la-Neuve, Belgium: Editions l'Harmattan.

_____. 1981b. *Maidens, Meal, and Money: Capitalism and the Domestic Community.* Cambridge: Cambridge University Press.

Miles, William F. S. 1994. *Hausaland Divided.* Ithaca, NY: Cornell University Press.

Ministère du Plan. 1993. *Les Arrondissements du Niger.* Niamey: République du Niger.

Monimart, Marie. 1989. *Femmes du Sahel: La Desertification au Quotidien.* Paris: Club du Sahel, Karthala, and OECD.

Mortimore, Michael. 1967. Land and population pressure in the KCSZ, northern Nigeria. *Advancement of Science* 23: 677–688.

_____. 1982. Framework for population mobility: The perception of opportunities in Nigeria. In *Redistribution of Population in Africa,* ed. J. Clarke and L. Kosinski. London: Heinemann.

_____. 1989. *Adapting to Drought.* Cambridge: Cambridge University Press.Munson, Patrick. 1986. Africa's prehistoric past. In *Africa,* ed. P. Martin and P. O'Meara. Bloomington: Indiana University Press.

Myers, Norman. 1993. *Ultimate Security: The Environmental Basis of Political Stability.* New York: Norton.

Newman, Paul, and Roxana Ma Newman. 1977. *Modern Hausa–English Dictionary.* Ibadan, Nigeria: University Press.

Ngaido, Tidiane. 1994. Natural resources management or decision making management: The tenure question in the Africare Gouré project. Unpublished report for Africare.

Nicholson, Sharon E. 1978. Climatic variations in the Sahel and other African regions during the past five centuries. *Journal of Arid Environments* 1: 3–24.

Nicolas, Guy. 1966. Essai sur les structures fondamentales de l'espace dans la cosmologie Hausa. *Journal de la Société des Africanistes* 36, 1: 65–107.

NRC (National Research Council). 1983. *Environmental Change in the West African Sahel.* Advisory Committee on the Sahel. Washington, DC: National Academy Press.

_____. 1992. *Global Environmental Change.* Washington, DC: National Academy Press.

Oberai, A. S., and H. K. M. Singh. 1980. Migration, remittances, and rural development: Findings of a case study in the Indian Punjab. *International Labour Review* 119, 2: 229–241.

_____. 1983. *Causes and Consequences of Internal Migration.* Delhi: Oxford University Press.

OECD (Organization for Economic Cooperation and Development). 1988. *The Sahel Facing the Future: Increasing Dependence or Structural Transformation.* Paris: OECD.

_____. 1994. *Development Co-Operation.* Paris: OECD.

Olivier de Sardan, Jean-Pierre. 1984. *Les Sociétés Songhay-Zarma (Niger, Mali): Chefs, Guerriers, Esclaves, Paysans.* Paris: Karthala.

Olofson, Harold. 1985. The Hausa wanderer and structural outsiderhood: An emic and etic analysis. In *Circulation in Third World Countries,* ed. R. M. Prothero and M. Chapman. London: Routledge.

Oucho, J., and W. T. S. Gould. 1993. Internal migration, urbanization, and population distribution. In *Demographic Change in Sub-Saharan Africa,* ed. Karen Foote, Kenneth Hill, and Linda Martin. Washington: National Academy Press.

Painter, Thomas M. 1985. Peasant migrations and rural transformations in Niger: A study of incorporation within a West African capitalist regional economy, c. 1875 to c. 1982. Ph.D. diss., SUNY Binghamton Department of Sociology.

_____. 1987. Making migrants: Zarma peasants in Niger, 1900–1920. In *African Population and Capitalism,* ed. D. D. Cordell and J. W. Gregory. Boulder, CO: Westview.

_____. 1992. *Migrations et SIDA en Afrique de l'Ouest: Etude des Migrants du Niger et du Mali en Côte d'Ivoire: Contexte Socio-economique.* CARE report. New York: CARE.

Panayotou, Theodore. 1994. The population, environment, and development nexus. In *Population and Development: Old Debates, New Conclusions,* ed. R. Cassen. Washington, DC: Overseas Development Council.

Panofsky, Hans E. 1961. *A Bibliography of Labor Migration in Africa South of the Sahara.* Evanston, IL: University Library, Northwestern University.

Pearce, Fred. 1992. Mirage of the shifting sands. *New Scientist,* December 12, 1992, 38–42.

Pehaut, Yves. 1970. L'Arachide au Niger. In Bibliothèque de l'Institut d'Etudes Politiques de Bordeaux, série Afrique noire, volume 1, Etudes d'Economie Africaine, 9–103. Bordeaux, France: Institut d'Etudes Politiques de Bordeaux.

Petersen, W. 1958. A general typology of migration. *American Sociological Review* 23, 2: 259.

Pingali, Prabhu. 1990. Institutional and environmental constraints to agricultural intensification. In *Rural Development and Population: Institutions and Policy,* supplement to volume 15, *Population and Development Review,* ed. G. McNicoll and M. Cain. New York: Oxford University Press.

Pingali, Prabhu, Y. Bigot, and H. P. Binswanger. 1987. *Agricultural Mechanization and the Evolution of Farming Systems in Sub-Saharan Africa.* Baltimore: Johns Hopkins University Press.

Pingali, Prabhu, and H. P. Binswanger. 1991. Population density and farming systems. In *Population, Food, and Rural Development,* ed. R. D. Lee et al. Oxford: Clarendon Press.

Pittin, René. 1984. Migration of women in Nigeria: The Hausa case. In *International Migration Review* 18, 4: 1293–1314.

Portes, Alejandro, and József Böröcz. 1989. Contemporary immigration: theoretical perspectives on its determinants and modes of incorporation. *International Migration Review* 23, 2: 606–630.

Portes, Alejandro, Manuel Castells, and Lauren A. Benton, eds. 1989. *The Informal Economy.* Baltimore: Johns Hopkins University Press.

Prothero, R. Mansell. 1957. Migratory labour from North-western Nigeria. *Africa* 27, 3: 251–261.

_____. 1959. Migrant labour from Sokoto Province. Unpublished manuscript. Kaduna.

Prothero, R. Mansell, and Murray Chapman, eds. 1985. *Circulation in Third World Countries.* London: Routledge.

Ravenstein, E. G. 1885. The laws of migration. *Journal of the Royal Statistical Society* 48, 2: 167–227.

Raynaut, Claude. 1977. Aspects Socio-économiques de la circulation de la nourriture dans un village Hausa. *Cahiers d'Etudes Africaines* 17: 569–597.

_____. 1988. Aspects of the problem of land concentration in Niger. In *Land and Society in Contemporary Africa*, ed. R. E. Downs and S. P. Reyna. Hanover, N.H.: University Press of New England.

Raynaut, Claude, ed. 1997. *Societies and Nature in the Sahel*. New York: Routledge.

Reenburg, Anette, and Bjarne Fog. 1995. The spatial pattern and dynamics of a Sahelian agro-ecosystem. *GeoJournal* 37, 4: 489–497.

République du Niger. 1990. *Migration, Urbanisation, Emploi Amenagement du Territoire*. Groupe 3. Comité Technique Interministeriel sur la Population. Project de politique de Population. Document de Base.

Rhodes, Steven L. 1991. Rethinking desertification: What do we know and what have we learned? *World Development* 19, 9: 1137–1143.

Richards, Paul. 1983. Ecological change and the politics of African land use. *African Studies Review* 26, 2: 1–72.

_____. 1985. *Indigenous Agricultural Revolution: Ecology and Food Production in West Africa*. London: Hutchinson.

Riddell, Barry J. 1980. African migration and regional disparities. In *Internal Migration Systems in the Developing World*, ed. R. N. Thomas and J. M. Hunter. Boston: G. K. Hall.

Roberts, Bryan. 1989. Employment structure, life cycle, and life chances: Formal and informal sectors in Guadalajara. In *The Informal Economy*, ed. A. Portes, M. Castells, and L. A. Benton. Baltimore: Johns Hopkins University Press.

Rochette, René M. 1988. Migration and settlement of new lands. In *Desertification Control and Renewable Resource Management in the Sahelian and Sudanian Zones of West Africa*, ed. F. Falloux and A. Mukendi. World Bank Technical Paper no. 70. Washington, DC: World Bank.

Rochette, René M., ed. 1989. *Le Sahel en Lutte Contre la Desertification*. Deutsche Gesellschaft für Technische Zusammenarbeit. Berlin: Verlag Josef Marglaf.

Rostow, Walt W. 1952. *The Process of Economic Growth*. New York: Norton.

Rouch, Jean. 1950. Les Sorkawa pêcheurs itinérant du moyen Niger. *Africa* 20, 1: 5–25.

_____. 1956. *Migrations au Ghana*. Paris: Société des Africanistes.

_____. 1961. Note sur l'importance des migrations pour le pays d'origin: Niger. CCTA/CSA, MIG (61) 18.

Salifou, André. 1971. *Le Damagaram; ou, Sultanat de Zinder au XIX Siècle*. Niamey: IRSH.

Schraeder, Peter J. 1997. France and the Great Game in Africa. *Current History* 96, 610: 206–211.

Schroeder, Richard. 1985. *Gender Vulnerability to Drought: A Case Study of the Hausa Social Environment*. Master's thesis, Department of Geography, University of Wisconsin–Madison.

Scott, James. 1976. *The Moral Economy of the Peasant: Rebellion and Subsistence in Southeast Asia*. New Haven: Yale University Press.

_____. 1985. *Weapons of the Weak: Everyday Forms of Peasant Resistance*. New Haven: Yale University Press.

Sen, Amartya. 1981. *Poverty and Famines*. Oxford: Clarendon Press.

Shaikh, Asif, and Michael McGahuey. 1994. Capitalizing on change: USAID's contribution to Niger's strategy for sustainable development. Draft USAID document.

Sidikou, Arouna Hamidou. 1978. Profession: Kourmizé. *Le Sahel*, March 14–15, 1978, 1182–1183.

Simon, Julian. 1981. *The Ultimate Resource*. Oxford: Martin Robertson.

Sivakumar, M. V. K. 1991. Climate change and implications for agriculture in Niger. *Climatic Change* 20, 4: 297–312.

Skeldon, Ronald. 1990. *Population Mobility in Developing Countries*. London: Belhaven.

Slim, Hugo, and Paul Thompson. 1995. *Listening for a Change: Oral Testimony and Community Development*. Philadelphia: New Society Publishers.

Smil, Vaclav. 1994. How many people can the earth feed? *Population and Development Review* 20, 2: 255–292.

Smith, M. G. 1962. Exchange and marketing among the Hausa. In *Markets in Africa*, ed. P. Bohannan and G. Dalton. Evanston, IL: Northwestern University Press.

_____. 1965. The Hausa of northern Nigeria. In *Peoples of Africa*, ed. J. L. Gibbs. Prospect Heights, IL: Waveland Press.

Société Générale des Techniques Hydro-Agricoles. 1964. Le Goulbi Nkaba et ses effluents: Mission de reconnaissance. Paris: Société Générale des Techniques Hydro-Agricoles (SOGETHA).

Speirs, Mike, and Ole Olsen. 1992. Indigenous integrated farming systems in the Sahel. World Bank Technical Paper no. 179. Washington, DC: World Bank.

Standing, Guy. 1981. Migration and modes of exploitation: Social origins of immobility and mobility. *Journal of Peasant Studies* 8: 173–211.

_____. 1982. *Conceptualizing Territorial Mobility in Low-Income Countries*. Geneva: International Labour Office.

Stark, Oded. 1991. *The Migration of Labor*. Cambridge, MA: Blackwell.

Stevens, William K. 1994. "Threat of encroaching deserts may be more myth than fact." *New York Times*, January 18, 1994, C1–C10.

Suhrke, Astri. 1991. Environmental change, population displacement, and acute conflict. Meeting paper. Ottawa: Project on Environmental Change and Acute Conflict.

Swindell, Kenneth. 1984. Farmers, traders, and labourers: Dry season migration from north-west Nigeria, 1900–33. *Africa* 54, 1: 3–19.

Taylor-Powell, Ellen. 1992. Social soundness analysis: Disaster preparedness and mitigation project. USAID Working Papers vol. 1.

Terraciano, Annmarie M. 1995. Contesting terrains: Tenure reform and the social dimensions of land conflict. Paper prepared for the Association of American Geographers Annual Meeting, March. Chicago, Illinois.

Thom, Derrick J. 1975. *The Niger-Nigeria Boundary, 1890–1906: A Study of Ethnic Frontiers and a Colonial Boundary*. Athens, Ohio: Ohio University Center for International Studies.

Thomas, David S. G. 1993. Sandstorm in a teacup? Understanding desertification. *The Geographical Journal* 159, 3: 318–331.

Thomas, David S. G., and Middleton, Thomas. 1994. *Desertification: Exploding the Myth.* Chichester, England: Wiley.

Thomas, J. J. 1992. *Informal Economic Activity.* Ann Arbor: University of Michigan Press.

Thomas, Stryk. 1992. A study of food security perceptions, Tillaberi Department. In *Perceptions of Famine and Food Insecurity in Rural Niger,* ed. Jane Hopkins and Ellen Taylor-Powell. USAID Working Papers, vol. 1. Washington, DC: USAID.

Tiffen, M., M. Mortimore, and F. Gichuki. 1994. *More People, Less Erosion.* New York: John Wiley and Sons.

Todaro, Michael P. 1969. A model of labor migration and urban unemployment in less developed countries. *American Economic Review* 59: 138–148.

Tucker, C., H. E. Dregne, and W. W. Newcomb. 1991. Expansion and contraction of the Sahara Desert from 1980 to 1990. *Science* 253: 299–301.

Turner, B. L. II, Goran Hyden, and Robert Kates, eds. 1993. *Population Growth and Agricultural Change in Africa.* Gainesville: University Press of Florida.

Turner, B. L. II, William B. Meyer, and David L. Skole. 1994. Global land-use/land-cover change: Towards an integrated study. *Ambio* 23, 1: 91–95.

Turner, B. L. II, R. H. Moss, and D. L. Skole. 1993. Relating land use and global land-cover change. IGBP Report no 24.

Turner, Matthew. 1993. Overstocking the range. *Economic Geography* 69, 4: 402–421.

UN (United Nations). 1979. *Demographic Yearbook.* New York: United Nations.

_____. 1995. *Demographic Yearbook.* New York: United Nations.

UNDP (United Nations Development Program). 1997. Poverty clock. January 17, 1996. *http://www.undp.org/undp/poverty/clock.htm* [1997, July 20].

UNEP (United Nations Environment Program). 1992. Status of desertification and implementation of the United Nations plan of action to combat desertification. UNEP/GCSS.III/3, Nairobi.

_____. 1993. *Climate Institute Programme on Environmental Refugees: Planning Document.* Oxford: The Climate Institute.

_____. 1995. *United Nations Convention to Combat Desertification in Those Countries Experiencing Serious Drought and/or Desertification, Particularly in Africa.* Nairobi: United Nations Environment Program.

UNHCR (United Nations High Commission for Refugees). 1995. *The State of the World's Refugees: In Search of Solutions.* New York: Oxford University Press.

USAID (United States Agency for International Development). 1992. Disaster preparedness and mitigation program: Program assistance approval document.

_____. 1994. Capitalizing on change. Draft working document, USAID Niger. Washington, DC: Agency for International Development.

van der Pol, Floris. 1994. *Soil Mining: An Unseen Contributor to Farm Income in Southern mali.* Netherlands: Royal Tropical Institute.

van Meter, Karl M. 1990. Sampling and cross-classification analysis in international social research. In *Comparative Methodology,* ed. E. Øyen. Newbury Park, CA: Sage Publications.

von Maydell, Hans-Jürgen. 1990. *Trees and Shrubs of the Sahel*. Weikersheim, Germany: Verlag Josef Margraf.

Wallerstein, Immanuel. 1974. *The Modern World System*. New York: Academic Press.

WALTPS (West African Long-Term Perspectives Study). 1994. *Description du peuplement de l'Afrique de l'Ouest*. WALTPS. Paris: OECD/CILSS. SAH/D(93) 415.

Warren, Andrew. 1993. Desertification as a global environmental issue. *GeoJournal* 31, 1: 11–14.

Watts, Michael J. 1983a. *Silent Violence*. Berkeley and Los Angeles: University of California Press.

_____. 1983b. On the poverty of theory: Natural hazards research in context. In *Interpretations of Calamity*, ed. K. Hewitt. Boston: Allen and Unwin.

_____. 1987. Drought, environment, and food security: Some reflections on peasants, pastoralists, and commoditization in dryland West Africa. In *Drought and Hunger in Africa: Denying Famine a Future*, ed. M. Glantz. New York: Cambridge University Press.

_____. 1989. The agrarian question in Africa: Debating the crisis. *Progress in Human Geography* 13, 1: 1–41.

Watts, Michael J., and Hans G. Bohle. 1993. The space of vulnerability: The causal structure of hunger and famine. *Progress in Human Geography* 17, 1: 43–67.

Williamson, Jeffrey. 1988. Migration and urbanization. In *Handbook of Development Economics*, volume 1, ed. H. Chenery and T. N. Srinivasan.

Wisner, Ben. 1993. Disaster vulnerability: Scale, power, and daily life. *GeoJournal* 30, 2: 127–140.

Wood, William B. 1994. Crossing the line: Geopolitics of international migration. In *Reordering the World: Geopolitical Perspectives on the Twenty-first Century*, ed. G. J. Demko and W. B. Wood. Boulder, CO: Westview Press.

Yo, Alvin Y. 1990. *Social Change and Development: Modernization, Dependency, and World-System Theories*. London: Sage.

Yusuf, Ahmed Beitallah. 1974. A reconsideration of urban conceptions: Hausa urbanization and the Hausa rural-urban continuum. *Urban Anthropology* 3: 200–221.

Zachariah, K. C., Julien Condé, N. K. Nair, Chike S. Okoye, Eugene K. Campbell, M. L. Srivastava, and Kenneth Swindell. 1980. *Demographic Aspects of Migration in West Africa*. Volume 1. World Bank Staff Working Paper no. 414. Washington, DC: World Bank.

Zolberg, Aristide, Astri Suhrke, and Sergio Aguayo. 1989. *Escape from Violence: Conflict and the Refugee Crisis in the Developing World*. New York: Oxford.

Index

DATE DUE
